Contributions to Management Science

For further volumes:
http://www.springer.com/series/1505

Pasi Tyrväinen • Oleksiy Mazhelis
Editors

Vertical Software Industry Evolution

Analysis of Telecom Operator Software

Physica-Verlag

Editors
Prof. Pasi Tyrväinen
University of Jyväskylä
Department of Computer Science
and Information Systems
Mattilanniemi 2
FI-40014 University of Jyväskylä
Finland
Pasi.Tyrvainen@jyu.fi

Dr. Oleksiy Mazhelis
University of Jyväskylä
Information Technology Research Institute
Mattilanniemi 2
FI-40014 University of Jyväskylä
Finland
Oleksiy.Mazhelis@jyu.fi

ISBN 978-3-7908-2805-4 e-ISBN 978-3-7908-2352-3
DOI: 10.1007/978-3-7908-2352-3
Springer Heidelberg Dordrecht London New York

Softcover reprint of the hardcover 1st edition 2009

Cover design: SPI Publishers Services

Printed on acid-free paper

Physica-Verlag is a brand of Springer-Verlag Berlin Heidelberg
Springer-Verlag is part of Springer Science+Business Media (www.springer.com)

Acknowledgements

The content of this book is mainly based on the research findings produced in an academia–industry collaboration project titled Software Market Analysis – Case Telecom Operator Products (SmarTop) in the Faculty of Information Technology at the University of Jyväskylä, Finland. The project was funded by the Finnish National Agency for Technologies and Innovation (Tekes), and sponsors for the project were Nokia (later Nokia Siemens Networks), Tecnomen, Compel, Nethawk and a teleoperator from the Finnet Group, MPY. The project took place in the framework of the Tekes technology program Verso (Vertical Software Solutions). It was conducted in collaboration with research partners from The Research Institute of the Finnish Economy (ETLA) where Prof. Olli Martikainen was leading the research, and with the Technology Business Research Centre (TBRC) at the Lappeenranta University of Technology, running a related project.

The authors would like to thank all representatives of the industrial and research partners we were collaborating with. We are also grateful to the companies (both telecom operators in Europe and Far East, and the OSS/BSS software providers) that kindly agreed to participate in the interviews conducted in the frame of the project; these interviews provided an invaluable insight into the processes ongoing in the OSS/BSS software market today. We would also like to thank the other organizations that provided complementary (quantitative and qualitative) data for this research: Dittberner Associates; Forrester Research; TeleManagement Forum.

We wish to express our gratitude to the international experts with whom fruitful seminars and workshops were held on various topics considered in the project, in particular to professor Jeffrey Funk (National University of Singapore), with whom the issues of industry (and technology) evolution were discussed, and to professor Hong-Mei Chen and professor Rick Kazman (University of Hawaii and Carnegie Mellon/SEI) for the collaboration on the topics of software architecture evaluation and service orientation in telecom industry. We also greatly appreciate the help of Steve Legrand in improving the language of the manuscript.

Preface

The telecom operator software market (OSS/BSS market) is analyzed in this book from the viewpoints of the business processes reflecting the telecom business, from the technology viewpoint visible in OSS/BSS interfaces, and from the market viewpoint as presented in market statistics. Chapter 1 presents the scientific background and Chapters 2 and 3 present a detailed analysis. The results have been condensed in Section 4.2 from the business perspective. It presents a default scenario containing both the main past events impacting the OSS/BSS markets and the expected future evolution of this vertical software market, which is created by matching the available market data against the model of vertical software industry evolution and other theories presented. Section 4.3 describes the method used, which can also be applied for other vertical software markets. This preface outlines the main results for the OSS/BSS market from the business perspective.

The OSS/BSS market is exceptional when compared to other markets due to the eTOM business process standardization effort driven by TM Forum. Standardizing the OSS/BSS processes, data model and a common set of interfaces for applications have been well appreciated (although not yet fully adopted) by telecom operators and especially by IT integrators.

Software related to network equipment has traditionally been vendor-specific. Later, network technology standardization has been very active and new standards have been created at a fast pace. Each new network technology with different performance characteristics required new software. Thus the number of network interfaces to be integrated to OSS/BSS software has been overwhelming prohibiting OSS/BSS software from reaching a steady market with a single dominating design.

Telecom operators, having invested heavily into previous generations of network infrastructure, are forced to maintain the legacy systems till the end of their investment demortalization period. Simultaneously they are forced to provide new services enabled by new technology both to new customers and to the existing customer base connected with old network technology. When the software product vendors have not been able to support a full set of legacy technologies, the large established operators have been forced to use IT integration services and in-house development much more than small communications service providers (CSPs), which have preferred the use of products from smaller vendors.

There have been various attempts to evolve the vertically integrated, technology-specific OSS/BSS software architecture to a vertically disintegrated, layered, network technology independent architecture. This has succeeded best at the customer interface layer with the aid of generic software vendors adapting their horizontal software for the requirements of CSPs'. At the service layer the attempts known as IMS, service delivery platform or framework have remained merely as approaches enforcing an engineering discipline while reducing engineering effort. Neither all-IP nor other approaches has succeeded in this at the network layer due to various reasons, such as technical problems with roaming, unwillingness of CSPs to cut the balance sheet value of their investments to old network technology and uncertainty of new business models replacing lost revenue on voice communication with added value services, advertisements and other new sources.

All these changes related to all-IP adoption will have a major impact on OSS/BSS software. Billing systems accounting for half of the OSS/BSS product revenue will become less necessary if monthly flat fees are commonly adopted. OSS/BSS software for customer identity management relying on phone number or SIM card identity will have to be revised, etc. If advertisements were to become the major income for consumer communication service providers, the operator segment might evolve towards the dual model of i) bit-pipe operators servicing enterprise, customers and advertisement agencies funding the monthly fees of consumers, and ii) specialized value added service operators providing service packages or individual services for the packagers. In addition, network equipment vendors might contribute to the first category by providing capacity in a form of managed services.

In most of these scenarios the number of OSS/BSS software vendors providing software related to the network layer is likely to decrease along with the decreased number of network equipment vendors. As the analysis of mergers and acquisitions shows, the number of software vendors seems to be decreasing although the vertical software market has not matured yet. As a general implication, the industry evolution life cycle and the vertical software market evolution seem to be only loosely coupled and the lack of software market growth causes market consolidation even in the middle of the vertical market evolution. However, the number of software vendors with products related to customer interaction is more difficult to predict as the border line between CSP software providers and the vast number of internet service software providers becomes more blurred due to the market convergence.

Contents

Contributors

Lauri Frank, Dept. of Comp. Sci. & IS, P.O. Box 35, FIN-40014 University of Jyväskylä, Finland, Lauri.Frank@jyu.fi

Eetu Luoma, Dept. of Comp. Sci. & IS, P.O. Box 35, FIN-40014 University of Jyväskylä, Finland, Eetu.Luoma@jyu.fi

Olli Martikainen, University of Oulu, Department of Information Processing Science, P.O. Box 3000, FIN-90014 University of Oulu, Finland
and The Research Institute of the Finnish Economy (ETLA), Lönnrotinkatu 4 B, FIN-00120, Helsinki, Finland,
olli.martikainen@etla.fi

Oleksiy Mazhelis, Information Technology Research Institute, P.O. Box 35, FIN-40014 University of Jyväskylä, Finland
Oleksiy.Mazhelis@jyu.fi

Mirja Pulkkinen, Information Technology Research Institute, P.O. Box 35, FIN-40014 University of Jyväskylä, Finland,
Mirja.Pulkkinen@jyu.fi

Kimmo Suojapelto, Technology Business Research Center, P.O BOX 20, 53851 Lappeenranta, Finland,
kimmo.suojapelto@formin.fi

Pasi Tyrväinen, Dept. of Comp. Sci. & IS, P.O. Box 35, FIN-40014 University of Jyväskylä, Finland, Pasi.Tyrvainen@jyu.fi

Jari Veijalainen, Dept. of Comp. Sci. & IS, P.O. Box 35, FIN-40014 University of Jyväskylä, Finland, Jari.Veijalainen@jyu.fi

Introduction

Pasi Tyrväinen and Mirja Pulkkinen

1 Overview

This book examines the evolution of vertical software industry, focusing on the industry of telecommunications software. In particular, this study analyzes the diffusion and vertical disintegration of the operations and business support systems (OSS/BSS) commonly deployed by contemporary telecom operators. Taking as a basis several models of industry evolution (technology diffusion, network externalities, evolution of the product, evolution of the vertical software industry), this book considers the current state of the telecommunications software market, analyzes the factors affecting further evolution of the OSS/BSS market, and describes scenarios of the market evolution in the future.

The current state of the telecom software market is considered in the book from the viewpoints of both telecom operators and software vendors. The analysis suggests that the market evolution has not yet reached its stage of maturity and large developments are still expected in near future. The analysis is based on both commercial databases on software market transactions and on interviews of operators in Europe and Far East. Both quantitative and qualitative methods are applied for the analysis.

The book is intended to decision makers in the CSP business as well as to those developing software for this rapidly evolving business sector. The aim is to do research that would satisfy academic standards while providing help in practical business problems for CSPs and software developers, i.e., for companies that are shaping the future of communications and sometimes found in one and the same enterprise. Seldom has any business been in such turmoil as the CSP business is today. This book opens up the possibility to approach analytically and factually the unknowns of the future of this business. For this purpose, Chap. 4 presents a

P. Tyrväinen (✉)
Department of Computer Science and IS, University of Jyväskylä, P.O. Box 35, 40014, Jyväskylä, Finland
e-mail: Pasi.Tyrvainen@jyu.fi

P. Tyrväinen and O. Mazhelis (eds.), *Vertical Software Industry Evolution*, Contributions to Management Science,
DOI 10.1007/978-3-7908-2352-3_1,

practical toolkit for managers for fast examination of the directions of the OSS/BSS software market. This is followed by a description about how this methodology was developed and instructions on how to apply the methods for another vertical software market.

2 Background and Goals

The content of this book is mainly based on the research findings produced in an academia–industry collaboration project titled Software Market Analysis – Case Telecom Operator Products (SmarTop) in the Faculty of Information Technology at the University of Jyväskylä, Finland.

The project was launched with the aim to analyze the status and evolution of a target vertical software industry. The target software industry chosen was software deployed by communications service providers (CSP). Due to convergence of communication services, the scope included fixed line, broadband, IP-TV, virtual and mobile communication service providers. The practical goal of the project was to analyze this market and the research goal was to develop a generic methodology for analyzing a vertical software market. The methodology employed was based on the model of vertical software market evolution resulting from previous research.

The target domain of telecom operator software used to be dominated by internal software development of national and regional operators as well as software developed by network equipment vendors. Due to introduction of new mobile communication technology and deregulation the operators have entered a phase of fierce competition. Since the start of the 2-year research undertaking (2007–2008), various technological developments have touched the CSP area of business. Firstly, during that time the accelerating diffusion of high capacity IP networks (broadband) has had an impact in communications in general all around the Western world, turning the interest of consumers and suppliers to the IP services potentially at the cost of telephony services that include both the old PSTN and mobile services using the evolving GSM and 3G technologies. Secondly, there exists a spread of wireless LANs, and the wireless broadband (e.g., WiMAX) technology as a swift way to provide access to IP services in areas with no cabling (copper, coaxial or fibre). Thirdly, the advanced mobile network technologies (such as HSPA) are evolving towards the next generations and enabling converged data, communications and media services on mobile terminals. Further, a broader variety of services, including diverse media and content offerings (e.g., video and TV on handsets), are becoming enabled with mobile networks and enhanced mobile terminals.

All this makes the targeted software market a research focus area of utmost interest. Looking into the service orientation in architectures and business models that is gaining ground in ICT applications, the CSP and the software markets seem to have a common future. The convergence of ICT and telephony, the mobile and IP services that can be extended to providing software as a service and cloud computing will lead to new business opportunities and business innovations.

The development of the OSS/BSS software market is considered from three complementary viewpoints: business process, technology, and software market.

Process. Software is deployed to automate and improve the execution of business processes. In a competitive market situation, standard level business process support is required to enter the market or retain a position in it. Advantage in competition can be achieved through improvement in the processes, unique services or quality outperforming the competition. Therefore, the research started out with a look into the current situation of business processes and the currently available software offering for CSP processes by comparing the software offerings to the TM Forums enhanced Telecomm Operations Map (eTOM).

Technology. Hardware technology innovation and evolution are complemented by developments in software. Hardware and software interfaces represent a crucial factor enabling new product and service development for new network technologies. Proprietary interfaces keep the control of related developments in the hands of the developer, whereas opening the interfaces for common use opens up market segments for a variety of applications. Therefore, considering software implementation in different phases of standardization, as well as considering the integration of systems using software developed by multiple companies are in the second focus area of the research project.

Software Market. The software market and the recent developments in it are naturally the third focus area. The market data being analyzed includes both the data supplied by software companies in this market and the data from telecom operator companies buying their software. The analysis of the CSP business is elaborated both by interviews on their software procurement and by analysing CSP business process improvement.

Some of the research questions addressed in the project and the book are:

What is the current state of the CSP software market evolution?

Does this market vertical comply with the findings about other software market verticals?

How and when will horizontalisation emerge in this vertical software market?

What strategies do the software vendors follow in this vertical?

Will vertical integration strategies of systems (including software) and vertical industry software solutions emerge? Or will the market rather be dominated by commodity hardware and software?

What will be the role of service oriented architectures in this market? Is it considered by the related enterprises as a market segment or as a technical platform with potential micro ecology / IPR protection strategies? Do software component markets evolve from service oriented architectures to service delivery platforms?

How to support telecom operators on "make or buy?" decisions by estimating development of specific market segments?

The distilled outcome of the research describes expected future scenarios for the vertical software market, and critical factors impacting the business and competitive positions.

3 Data Sources Used

One of the challenges in analysing an emerging or an evolving market is the availability of sufficient and reliable data. In this book a wide variety of heterogeneous data sources are used, varying from public and commercial market research reports and databases to interviews with domain experts. Both qualitative and quantitative data are employed. Different sources of data were often required when analyzing the vertical software market from the viewpoint of telecom processes, markets, and related technologies.

One of the main sources of quantitative data used in the book is a commercial data set (coined OSS/BSS KnowledgeBase) acquired from Dittberner Associates (www.dittberner.com), an international market research and consulting firm with focus on the telecommunications industry. It offers surveys and related data on the OSS/BSS industry development. The data used includes the OSS/BSS vendor revenues from the years 2000, 2002, 2005, and 2006, and it covers 187 companies of the approximated total of 250 companies present in the OSS/BSS industry. Therefore, the data excludes the smallest and only regionally operating vendors. This is the only market research data available to the authors which provides vendor specific information for the OSS/BSS. The data is based on the vendors' self evaluation.

In addition to the data by Dittberner Associates, commercial data sets by other market research firms (ThomsonOne, Amadeus, ArcChart, Forrester, etc.) were used. Whenever possible, public sources of information, such as TMF, ITU, CDMA2000, GSMA, LightReading, and OSS Observer, were used as well. When necessary, additional information was obtained through the Annual Reports of the OSS/BSS companies and CSPs.

Finally, in order to obtain a deeper insight into the developments in the OSS/BSS domain, available commercial and public datasets were augmented with the qualitative data gathered in the course of interviews with representatives of CSPs and telecom service providers in Europe and in the Far East. Besides, valuable information was obtained by consulting visiting domain experts.

4 Outline of the Book

The remainder of the book is organized into four parts as follows:

Part I. Evolving Software Markets

The first chapter starts by describing the birth of the software industry as a spin-off of computer manufacturing industry and outlines the role of vertical software industries in the intersections of the primary software industry and other industries purchasing specialized software systems for the needs of their core business processes. The three perspectives of analysis are introduced: vertical industry business processes, technologies used through software interfaces and software market.

The second chapter reviews theories related to vertical software industry evolution. These include innovation diffusion theory, evolution of the product during diffusion, industry evolution stages, and vertical and horizontal industry evolution.

Third chapter represents the model for evolution of a vertical software industry. In this model a vertical software industry includes both the in-house software development within the vertical industry enterprises and the vertical software market, the latter of which means the sales of software products and services from the primary software industry companies to the vertical industry enterprises. During the evolution of a vertical software industry the software development shifts from in-house development to purchase of software products and related services, which is visible in growing vertical software market volume.

The last chapter gives an overview of the telecom operator software market referred to as the OSS/BSS market. This includes operations support systems (OSS) and business support systems (BSS) used by telecom operators. The core business processes of operators are described using the eTOM framework developed by TM Forum. A consolidated set of OSS/BSS market segments is presented and mapped against the eTOM process framework. Basic data of the market segments presented includes volumes and growth rates of the segments as well as a number of OSS/BSS vendors in each segment. Vendors' strategies are also classified according to their breath of operations and their focus on the telecom market.

Part II. Current State of the Telecom Software Market

The part consists of three chapters which describe the role of software in optimizing the business processes of a CSP and consider the diffusion of the OSS/BSS software:

First chapter focuses on the role of software in optimizing the business processes of a CSP. As explained in the chapter, investments in information technology may improve productivity only when they are connected to organizational changes. Second chapter uses the data on software spending by CSPs to analyze the effect of software investments on CSP performance.

Third chapter discusses "make or buy" considerations by CSPs, based on interviews conducted with CSPs in Europe and Far East. The rapid change in telecom business environment taking place in China is also discussed.

Fourth chapter considers the current state of the telecom software market. In this chapter, the data by Dittberner Associates is employed in order to match the telecom software market evolution against the model for evolution of a vertical software industry, Rogers' innovation diffusion model, and Moore's technology adoption lifecycle. The chapter studies also correlations between OSS/BSS software market indicators, analyzes market concentration, studies strategies adopted by the software vendors, and considers the OSS/BSS software market evolution through alternative merger and acquisition strategies.

Part III. Issues Affecting Telecom Software Market Evolution

The part analyzes factors which are specific to software industry and telecom operator software market. These factors may potentially halt the development of the vertical (OSS/BSS) software market or change the direction of the development.

Chapter "Vertical Integration due to Software Systems' Modularity" analyzes the role of system modularity in predicting the integration of software subsystems within vendors' offerings. It seems that the data flows in the main business processes (order fulfilment, billing and assurance) enforce individual software systems to interface with other systems within the same process. The hypothesis analyzed is that this inherent connection enforces the software to be preintegrated in order to match the processes, to enable innovation, and to avoid integration costs repeated for each customer installation.

Chapter "Vertical Integration due to Small Market Size and High Product Development and Integration Costs" compares the costs of in-house software development against the costs of acquiring and integrating software provided by an independent software vendor. The hypothesis analyzed here is that a small number of customers sets a limit to the maximum number of software product vendors that can survive in that market segment. In an extreme case a limited market does not enable any software vendor to develop a product for the segment due to high development costs.

Chapter "Market Polarization due to Difference in Interface Implementation Efforts" analyzes the impact of continuously emerging new interfaces to network technologies due to new standards and new product generations. One of the predicted consequences is splitting of the OSS/BSS market into a submarket of large incumbent operators served by large software vendors providing knowledge for a large set of network interfaces and a submarket of commencing operators served by commencing software vendors adopting only standard network technologies supported by inexpensive software products.

The last chapter entitled "Barriers to Horizontal Development" analyzes the impact of patents, licenses, and their distribution to the market evolution.

Part IV. Future Scenarios for the Telecom Operator Software Market and Their Evaluation

Part IV outlines future scenarios for the OSS/BSS software market. The first chapter uses the Delphi method to evaluate and prioritize 16 most interesting miniscenarios of market evolution derived from the previous parts.

The second chapter presents a Manager's Toolbox for analysis of the OSS/BSS market. It represents a base-line scenario containing the past and present major events impacting the OSS/BSS market. The most relevant miniscenarios of the first chapter are attached to the base-line scenario as alternative elements of the base-line scenario. For each miniscenario, the toolbox presents the logic behind the miniscenario and threshold metrics for following the possible realization of the miniscenario.

The last chapter provides guidelines for developing further miniscenarios for the OSS/BSS market. This method can also be used for analyzing another vertical software industry (such as retail banking software market or automotive industry software market) and for creating a manager's toolbox for it.

5 Related Publications

This book, containing contributions from numerous researchers, summarizes the main results of the SmarTop project. Parts of the work have been published earlier in scientific forums, such as IEEE, IFIP, and ICETE. The related publications produced in earlier phases of the research are as follows:

Part I

- First chapter: (Tyrvainen et al. 2008)
- Third chapter: (Tyrvainen et al. 2008, 2004)
- Fourth chapter: (Frank et al. 2007)

Part II

- Second chapter: (Mazhelis et al. 2008a)
- Third chapter: (Pulkkinen et al. 2008)
- Fourth chapter: (Tyrväinen and Frank 2008; Frank and Luoma 2008; Mazhelis et al. 2008a)

Part III

- First chapter: (Mazhelis et al. 2008c)
- Second chapter: (Mazhelis et al. 2008d)
- Third chapter: (Mazhelis et al. 2008b)

Part I
Evolving Software Markets

Vertical Software Industries

Pasi Tyrväinen

The *software industry* (SWI) was born only a few decades ago, and is still relatively young compared to other industries. Its origins can be traced down to the decisions to unbundle some secondary parts of IBM computer development to independent software companies (Campbell-Kelly 2004). The software development was then considered a secondary activity serving the core business, which was at that time computer hardware manufacturing. This focus on hardware is still visible in the brand names of both the global information and communication technology (ICT) giants, such as IBM, and the major professional and scientific establishments of the field, such as the Association for Computing Machinery (ACM).

Gradually, however, software industry has established its position as an independent industry. Also, software business has become a specialized focus of scientific research. First, this research targeted the core of the emerging industry, i.e., independent software vendors (ISV) developing and selling independent software products and software services – these companies are often referred to as the *primary software industry* (BMBF 2000). The term *secondary software industry* refers to software business performed by companies that focus on another industry but utilize software as a part of their products or services. At different times, different vertical industries represented the key host industries for the secondary software industry: the automation industry in the 1970s, the electronics industry in the 1980s, and the telecommunication industry in the 1990s. At the moment, e.g., the automotive and to some extent the aerospace industries are the centre of attention for booming secondary software businesses.

P. Tyrväinen
Department of Computer Science and IS, University of Jyväskylä, P.O. Box 35, 40014 Jyväskylä, Finland
e-mail: Pasi.Tyrvainen@jyu.fi

P. Tyrväinen and O. Mazhelis (eds.), *Vertical Software Industry Evolution*, Contributions to Management Science,
DOI 10.1007/978-3-7908-2352-3_2, © Springer-Verlag Berlin Heidelberg 2009

1 Role of Vertical Software Industries in Global Software Business

In the early years of the twenty-first century, the primary software industry has been a global business with over 600 billion USD annual revenues, out of which about 200 billion USD came from software products, and over 400 billion USD from services (see Table 1). This means that roughly one-third of the revenue was produced by sales of licenses and two-thirds with IT services. Together, these two represent slightly less than a third of the total ICT market.

According to EITO (EITO 2004) figures of the software market, out of the 200 billion euros spent on software products annually, roughly one-half is spent on applications, one-third on software infrastructure software, and less than one-fifth on development tools. Furthermore, out of the applications, the software intended for some vertical market amounts to almost a half of the spending, the same as ERP software and other horizontal software, while less than a tenth of software products were sold to private consumers (Colecchia et al. 2002). Industry specific figures show also that software product purchases of the financial industry amount to three times the volume of the consumer market sales. The share of the private consumer market is also extremely small as compared to the over 400 billion euros software service market.

These global figures may hide the impact of strong vertical industries in the global software market. By a *vertical software industry*, we refer to the primary and secondary software industry organizations producing software for the needs of a vertical industry, such as construction or financial industry. For example, banks and other financial institutions alone represent close to 30% of the software market (see Table 1) with their 180 billion euros annual purchases from the primary software industry. In other words, the main volume of the primary software industry sales recorded as the software market comes from the large vertical businesses. This means that the financial industry and other strong verticals drive new software development of the primary SWI to the directions beneficial for these industries. In addition, several enterprises within the financial industry have internal software

Table 1 Primary software industry revenue as a part of the ICT industry. Amounts are rounded to billions of euros (EITO 2004; OECD 2002[1])

	Global	Europe	Financial industry
Software products	200	65	20
IT services	400+	120	160
Equipment		150	
Communication service		250	
ICT total	2200	600	400+

[1]OECD IT Outlook 2004 Available at http://www.oecd.org/dataoecd/22/18/37620123.pdf. See also "Measuring the Information Economy 2002" available at http://www.oecd.org/document/5/0,3343,en_2649_34449_2765701_1_1_1_1,00.html

development activity, which is not visible in the market figures representing only the purchases from the primary software industry enterprises. Another interesting observation from Table 1 is that in general the ratio of software product spending versus IT service spending is 1 to 2, while in the financial industry it is 1 to 8. This reflects an emphasis on tailored systems and process consulting as compared to a general tendency to use more standard product packages.

Figure 1 represents the overall settings, in which the primary and the secondary software enterprises interact. Primary software industry companies, placed on the left in the figure, run software business as their main activity, though they may have units with another business orientation. In the NACE categorization used by OECD and EITO, these companies are categorized mostly under category 72.2. Companies with other categorizations often have software-oriented business units or their operations include software development, sales and consulting related to their products. These companies are placed on the right hand side in the figure as the secondary software industry.

The official NACE-based statistics on software production provides only a partial picture of the software sales: the statistics reflects the total sales of the 72.2 category software companies (including also nonsoftware sales), while software sales of the software business units in the secondary software industry are excluded. Meanwhile, according to former EU Commissar Erkki Liikanen, in the EU, 70% of software was developed by the vertical industries, such as car manufacturers, while only 30% was produced by primary software industry enterprises.

This means that the software market statistics represent only the revenue of the primary SWI enterprises while the total volume of a vertical software industry includes also the software development activity of the secondary SWI enterprises, which often exceeds the primary software industry volume.

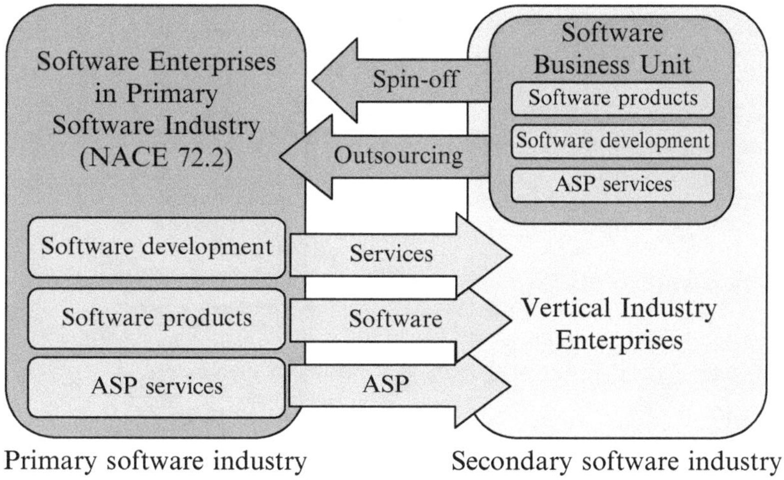

Fig. 1 Interaction between enterprises in the primary software industry and other industries

2 Three Perspectives Needed for Success of Vertical Software

New primary software companies may be born both from academic environments and technology developers, as well as from the business units of secondary software enterprises (Chesbrough et al. 2002; Giarratana 2004; Romijn et al. 2002). Both establishment of spin-off companies and outsourcing of software development tasks move personnel and business activities from vertical industries to the primary software industry.

The key factors impacting the success of software products aimed at a vertical market can be viewed from the three perspectives of Fig. 2, the technologies, the vertical industry processes and the position of the software enterprise in the market. All these three viewpoints impact, in many ways, the software to be implemented for automating a specific process.

From the vertical industry viewpoint, the processes or systems to be automated using software set both functional and nonfunctional requirements for the system.

Considering the functional requirements, it is clear that any piece of software should be customer-oriented (McHugh 1999) in the sense of satisfying certain information processing needs. This requires expertise from the target processes or systems. Bell (1995), for example, emphasizes the importance of targeting niche markets, industry specific conditions, and relationships with important customers in creating this expertise. In practice, expertise may initially be realized in the form of the functionality of a unique piece of software.

If the logic of a business process cannot be determined in advance, some parts of the operation have to be performed by a human and the level of automation reached through a software automating the business process will remain limited. In case the

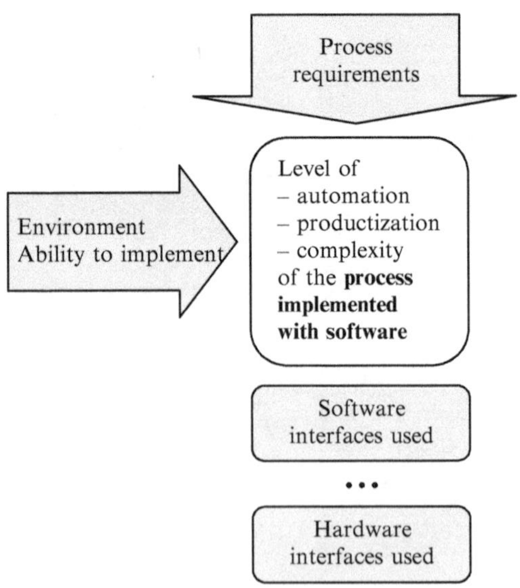

Fig. 2 Factors impacting the success of software solution in vertical markets

processes vary from an enterprise to another, the process cannot be fully productized and some tailoring, configuring and interfacing of the software will be needed for each enterprise adopting the software. Higher complexity of processes also requires higher complexity of software to automate it and thus requires also greater effort from the software vendor to implement it. The effort needed depends also on the availability of standardized enabling technologies with which the solution has been developed.

From the technology viewpoint, the major constraints are set by the interfaces through which the software interacts with other processes or systems. In the worst case, the software has to interface with proprietary systems through proprietary interfaces. In this case the software developer goes through an expensive process of acquiring information on the interfaces, may need specialized resources and personnel with expensive training, or the interfacing may not be possible at all. In the best case there exist an easy-to-interface system encapsulating the knowledge on how to access this hardware, or all the interfaces needed by the software are common and openly available on the marketplace. In other words, not only the required information processing functionalities and the technologies, but also system interfaces should become standardized (Northrop 2002).

Often, generic software products, which are being deployed across industries also have their origins in the vertical markets, from where they have, through a process of horizontalization, been adopted and adapted for general use. For example, manufacturing resource planning systems were initially developed as tailored solutions. Later on their scope was extended by the first adopters to other functions of the organization. Now such systems are widely adopted as generic enterprise resource planning (ERP) software systems also by other industries.

This phenomenon can be seen from the perspective of diffusion and adoption of enabling technologies. According to the diffusion models, software technology is originally technically embedded into, or at least business-wise bundled with hardware technologies. These diffusion models, as well as other theoretical models related to explaining and/or predicting the evolution in a vertical software industry are discussed next.

Theories Related to Vertical Software Industry Evolution

Lauri Frank

1 Innovation Diffusion Theory

The word innovation often refers to a new product, usually to an industrial or technical invention, although actually all kinds of new ideas should be considered. Following Rogers (2003, p. 12), "[a]n innovation is an idea, practice, or object that is perceived as new by an individual or other unit of adoption." The birth of a new industry can be considered to occur as this new innovation is introduced to the markets by the first manufacturing firm. At this point the potential customers evaluate whether it is worth to adopt the product or not.

Innovation diffusion theory studies the spread of an innovation within a social system (see, e.g., Rogers 2003). The basic aim is to answer the question of why an innovation – an improvement – is not adopted immediately, i.e., why its diffusion takes time. One of the common answers is that not everyone knows of the innovation right away, others add, for example, that gained advantages or profits vary between potential adopters.

There are several dichotomous classifications of innovations, trying to separate innovations into two categories by their nature. Some of such dichotomous divisions are, for example, the division of innovations into incremental and radical innovations (e.g., Freeman 1994), into sustaining and disruptive (e.g., Bower and Christensen 1995), or into continuous and discontinuous innovations (e.g., Tushman and Anderson 1986). Even though the category names differ, these divisions seem to have the same basic aim: An innovation may be a totally new innovation (e.g., product), but it also could be an improvement to an existing innovation. In practice, however, it is often hard to place an innovation into one of these two categories as it might have some characteristics from both.

L. Frank
Department of Computer Science and IS, University of Jyväskylä, P.O. Box 35, 40014 Jyväskylä, Finland
e-mail: Lauri.Frank@jyu.fi

P. Tyrväinen and O. Mazhelis (eds.), *Vertical Software Industry Evolution*, Contributions to Management Science,
DOI 10.1007/978-3-7908-2352-3_3,

Quantitative modeling of the time path of an innovation's diffusion has been traced back to the early 1940s: In their study, Ryan and Gross (1943) used the cumulative normal curve to explain the sigmoid shape of the diffusion of hybrid corn. In later studies, Griliches (1957) and Mansfield (1961) implemented analogies from theories of epidemiological disease spread on innovation diffusion models. Following this idea, information about an innovation would spread like a disease, by contagion from one carrier to another. Thus the diffusion of that innovation would reach a sigmoid shape on the macro level: In the beginning the diffusion would be slow, in the mid-phase fast, and in the end slow again (see Fig. 1). These three phases were named by Griliches (1957) as the origin, diffusion and saturation phases – but they are also commonly known as the innovation, growth and maturity phases. Because of the analogy to the spread of diseases, these models are also known as "epidemic" diffusion models.

Mathematically the epidemic theory is in line with the logistic growth model developed by P.-F. van Verhulst in 1848 for modeling population growth. The logistic model has been successfully applied in several technology diffusion studies, for example to the diffusion of mobile communications (e.g., Gruber and Verboven 2001a; Frank 2004). Instead of deriving the logistic model from the epidemic theory, it may also be derived from network effects, or from a demand function (see Wang 2006). In marketing, it is common to use the Bass (1969) diffusion model, which considers epidemic diffusion caused by "internal influence" (information exchange between adopters), and adds "external influence" (information from external sources) to it.

As mentioned above, several alternatives to the idea of passive information contagion behind an innovation's adoption and diffusion have been proposed. In the epidemic diffusion model a potential adopter would adopt a new innovation immediately upon receiving information about it. As the information about the innovation spreads by contagion, the diffusion is bound to take time.

Other theories assume, for example, that for some reasons the benefit or utility of potential adopters vary, and thus they adopt at different points of time. One of the

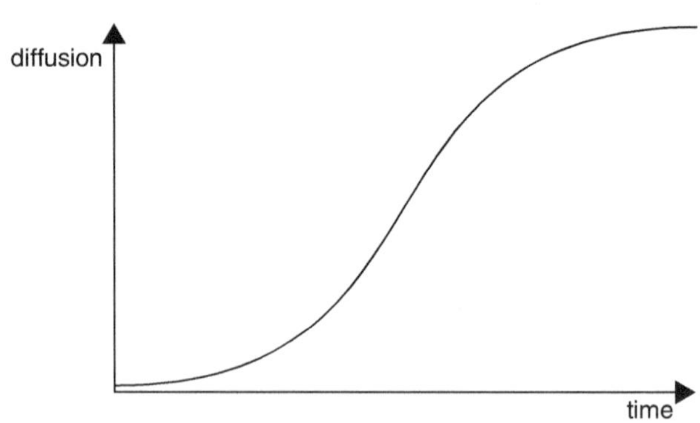

Fig. 1 The S-shaped innovation diffusion curve (see, for example, Rogers 2003)

proposed reasons for the varying benefits derived is the presence of network externalities or effects.

Economides (1991) states that network effects occur when "The buyer of the last unit of a good has a higher benefit than the buyer of the first because the sale of the earlier units has created some benefits in a related dimension" – i.e., the number of existing users (the installed base) affects the benefit of a new adopter. Positive network effects have been noted to influence the adoption and diffusion of many technological innovations (e.g., Church and Gandal 1993; Katz and Shapiro 1985; Witt 1997). Network externalities have also been seen as complicating the forecasting of the adoption success in telecommunications services (Schoder 2000). Network externalities can be seen to lead to the above mentioned logistic diffusion model, where the diffusion is also related to the number of existing users.

Network effects cause the presence of a critical mass point in the diffusion of an innovation: In the early phase of the diffusion the innovation does not seem attractive to potential adopters, because there are only few other users. If the critical mass is not reached, the diffusion will fail. On the other hand, if the critical mass point is reached, the innovation diffuses successfully. The critical mass point in the diffusion process is generally expected to occur approximately between a 10% and 20% diffusion level (Rogers 2003; Valente 1995).

Rogers (2003) also presents a model of the adoption process, which implies that the adoption of a new innovation is not a simply choice at a certain point of time, but usually a lengthy decision process which takes time. Organizations are considered to differ from consumers as adopters of innovations. For example, Cooper et al. (1990) analyze technology diffusion within an organization from the perspective of the end-user organization, i.e., the vertical industry enterprise. Their analysis is based on the work of Lewin (1947) and Rogers (1983). They found that the adoption process starts with initiation and adoption of the technology, continues with adapting it to the organization use, and continues after acceptance with routinization and infusion.

2 Evolution of the Product During Diffusion

The derivate of the above presented S-shaped innovation diffusion curve represents the number of technology adopters in a period of time. This bell-shaped adoption curve has its maximum in the middle, when the number of adopters per time unit is the highest. In addition to Rogers (2003), also Moore (1995) uses this curve to categorize technology adopters into five categories during a technology adoption life cycle (see Fig. 2). In his categorization, Innovators and Early Adopters are more willing to adopt a technology product. This alone does not solve the customer's problem, as it might require lots of complementary customer service. For example, Late Majority customers require the availability of a "Whole Product," which is "the minimum set of products and services needed for customers in the target segment to achieve the value proposition promised (Moore 1995, p. 165)."

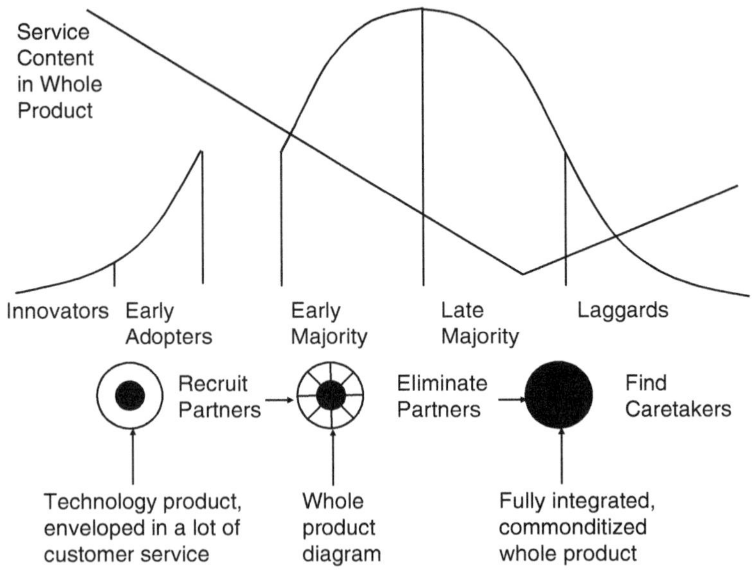

Fig. 2 The evolution of the "whole product" and share of services (Moore 1995, p.155) along with the technology adoption life cycle (Moore 1995, p. 14)

From Moore's viewpoint, gaining critical mass is needed for crossing a chasm between Innovators and Early Majority. The critical mass is usually obtained with the help of partners implementing complementary services for the technology product of the primary innovator company.

The concept of the whole product can easily be explained. In a software market, the whole product contains typically a core software product provided by a software product company and complementary integration services provided by, for example, an IT consulting company. For example, in the early phases of ERP technology evolution the product licenses of SAP accounted for about 20% of the sales price of a customer delivery (Hoch 1999) while services of IBM, Accenture and other IT integrators accounted for about 80% of the delivery. The price erodes along with the reducing volume of services, until in the late phases of the life-cycle the product vendors try to increase their sales volume by increasing service revenues. This development is well in line with the view of Moore presented in Fig. 2.

Along the same lines, the life-cycle and growth models proposed by Greiner (1972), Churchill and Lewis (1983), Scott and Bruce (1987), Nambisan (2002), Kazanjian (1988), and McHugh (1999) observe the life cycle of companies in general or high-technology companies, from the perspective of the software company rather than from the perspective of the status of some vertical market. However, Greiner (1972) states that the growth of a company is often related to the market environment and stage of the industry. From our perspective this is essential, since the development of software businesses servicing specific vertical markets depends both on the status of the primary software industry and the vertical market. Next we shall look at the industry evolution models.

3 Industry Evolution Stages

An industry can be considered to evolve like an organism: it is born, it lives through different phases, and finally it dies. The industry lifecycle theory considers the evolution of a new industry from the producer's viewpoint. It hypothesizes how a number of producers – who evaluate the viability of producing something – evolves through time. In step with this industry evolution process, the industry's market structure also changes.

The industry lifecycle theory has its origins in the seminal work by Gort and Klepper (1982). They constituted a theory of the evolution of new product industries, basing their effort on evidence from 46 product histories. The net entry rates of these product industries (the changes in the number of producers) were classified into five industry evolution stages through which most new products were thought to pass. Figure 3 illustrates these five lifecycle stages:

1. The introduction of a new commercial product by its first producer (or producers).
2. A period of rapid growth in the number of producers.
3. A period where the number of new entrants is about equal with the number of exiting producers.
4. A period of negative net entry, where more firms exit than enter the industry. This phase is also called the industry "shakeout" stage.
5. A second period where the number of entrants approximately equals those who exit. This is considered as the maturity phase of the industry.

Gort and Klepper (1982) assumed the entry rate to be determined by the potential returns to a firm, more specifically by the number of innovations from external sources, the accumulated stock of experience of incumbent producers, and the profit of incumbent producers. Agarwal (1998) has later proposed a quadratic formula for modeling the evolution of the number of firms.

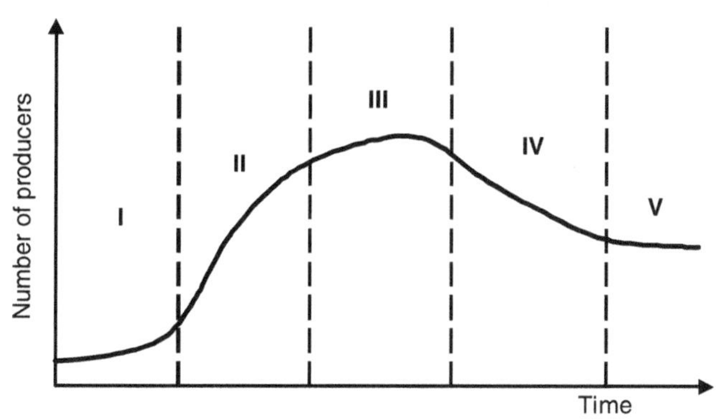

Fig. 3 Industry lifecycle stages (Dinlersoz 2005)

Some studies have combined the five stages proposed in the original model by Gort and Klepper (1982). Agarwal and Audretsch (1999) reduce the five industry lifecycle stages into two: The formative years when entrants try new ideas, and the mature period with a standardized product. These two periods are characterized by a positive and negative net entry, respectively. They find that a larger firm startup size correlates positively with its survival probability in an uncertain environment – the formative years of an industry. However, in the mature period there seems to be no relationship between firm size and survival. This suggests that in a certain environment and with a standardized technology, small startup firms may overcome the disadvantages of their size by a niche strategy. It is thought that a firm may gain competitive advantage over other firms by either choosing a price leadership or a specialization strategy: A firm may try to produce a mass product at a price lower than that of other firms, it may provide something unique, or it may try to focus on a niche market (see Porter 1985). An empirical study shows the actual strategies of firms operating in the OSS/BSS industry sectors include all the above mentioned strategies (Frank et al. 2007).

Dinlersoz and Hernández-Murillo (2005) consider three phases in an industry's lifecycle: The initial phase where many firms enter the industry, the second phase where the number of firms declines sharply, and the third phase with a stable number of surviving and successful firms. They illustrate this lifecycle with the development in the number of FM radio stations, and subsequently study the evolution of retail e-commerce. They note that the shakeout lasted only for a few months in the internet industry, whereas the usual average in manufacturing industries is several years or even decades. Easy access to the industry also caused a much more rapid initial phase of entry. They also conclude that an Internet retailer dominating one product category seems to be more capable of expanding also into other product categories.

The above theories and studies illustrate the evolution of the supply side. The supply and demand of a market are interconnected, as confirmed also by Wang (2006) who provides a hypothesized relationship between the supply side tracking industry lifecycle theory and the demand side focused innovation diffusion theory: He proposes that the decrease in demand in the maturity phase of an innovation's diffusion actually affects the number of firms of that industry. Thus demand characteristics should be taken into account when analyzing industries' lifecycles. In his results, he shows that faster technological learning, higher mean income and larger market size cause a faster diffusion process and subsequently an earlier industry shakeout.

4 Vertical and Horizontal Industry Evolution

Macher and Mowery (2004) have studied the evolution of the structure of industries. They look at the vertical specialization in three mature but high R&D level industries: The chemicals, computers and semiconductor industries. They define vertical specialization as "[. . .] the restructuring of industry-wide value chains, such that different stages of the development, production, and marketing processes are controlled by different firms, rather than being vertically integrated within the boundaries of individual firms."

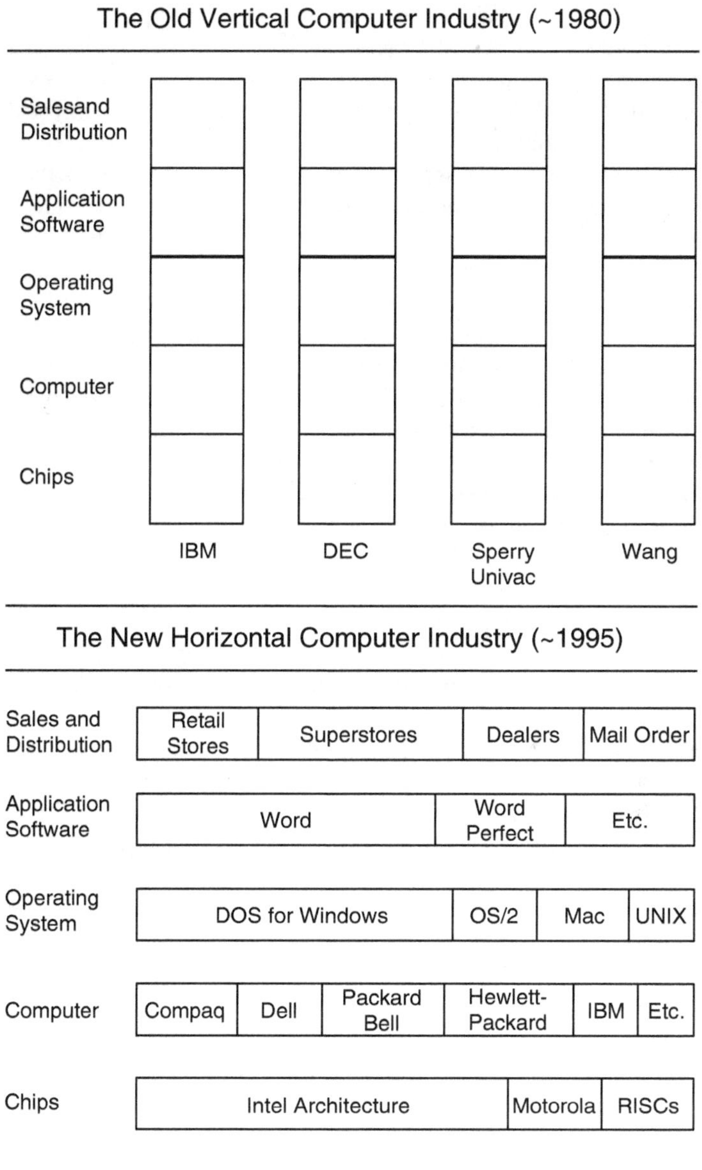

Fig. 4 The old and new structure of the computer industry (Bresnahan 1998)

Bresnahan (1998) provides an example of vertical specialization in his analysis of the competitive nature of the computer industry. His presentations of the structural change in the computer industry show that the structure has changed from vertically integrated firms to a vertically disintegrated or specialized structure (see Fig. 4).

Figure 4 shows how the computer industry consisted earlier of a couple firms which ruled the whole value chain, thus proving end-to-end solutions to customers. A more open architecture enabled vertical disintegration in the computer industry: since the mid-1990 s specialized firms have handled the provision of hardware, computers, operating systems, applications and sales. Bresnahan (1998) notes that vertical specialization of the computer industry lead to vertical competition or co-opetition, where the manufacturers of complementary products compete. Bresnahan and Greenstein (1999) further report that the competition within the computer industry is between platforms and not directly between firms, and that the vertical disintegration of the computer industry has lead to an intensified technological competition as new platforms have been able to challenge older platforms.

Usually vertical specialization is seen to lead to a "commoditization" of activities, which again attracts new entrants to the market and intensifies competition. It is often noted to emerge as entry barriers of an industry's segment are reduced (Macher and Mowery 2004). Stiegler (1951, cited by Macher and Mowery 2004), claims vertical specialization to occur in the early phase of an industry's lifecycle as entry firms start to produce those components which have "declining unit costs and a substantial minimum efficient scale of production." Re-integration would be common in the mature phase of an industry's lifecycle as vertically specialized firms are no longer supported by the market. It has been shown that often there are development cycles in an industry's structure, where vertical specialization (disintegration) and re-integration follow each other. For example, Macher and Mowery (2004) report vertical specialization (or vertical disintegration) in the computers and semiconductors industries, but already find evidence of vertical re-integration in the chemicals industry.

Macher and Mowery (2004) also provide an extensive review of studies looking at the reasons for vertical integration and specialization at the firm level. Reasons affecting firm behavior include, for example, transaction cost analysis (TCA) and the knowledge based view of the firm (KBV), which support vertical integration of firms. Following TCA, vertical integration provides a safeguard for the possible opportunistic behavior of trading partners. The KBV stresses, e.g., the increased efficiency caused by the common codes of conduct within the same firm. On the other hand, a vertically integrated industry may be inflexible in adapting to new specialized activities. This may lead to an outsourcing of such activities – as they are not the core of the industry's business – and further to vertical specialization. Vertically specialized firms are seen to be better in innovating new components, whereas vertically integrated firms handle changes of the whole system more efficiently.

Model for Evolution of a Vertical Software Industry

Pasi Tyrväinen

1 Phases of Evolution

The interaction of the primary and secondary software industries appears to follow a common pattern, which has been described in the model for evolution of a vertical software industry (Tyrväinen et al. 2008). The evolution is seen as iterating through several phases, shown in Fig. 1 and described below.

The first phase of the model is called the *Innovation phase*, where the vertical industry enterprises seeking to gain competitive advantage implement their core business processes with some innovative software developed in-house. In service-oriented businesses this means automating manual work with information systems and in companies developing and producing hardware products this means implementing most of the software needed to use their equipment by in-house R&D personnel. In both cases the vertical industry enterprise invests increasingly in ICT and the number of software personnel tends to grow gradually. In case the enterprise has to cut back on software workforce during economical downturn this can be done by relocating these employees to other duties in which they can perform well due to their industry competence. Some of the software development professionals may also become self-employed by establishing micro ventures and capitalizing on their expertise with new types of systems, thus serving as agents of technology diffusion.

In the second phase called *Productization and standardization phase*, the companies in the vertical industry improve their enterprise-specific software and copy best practices of their competitors to their in-house software thus standardizing the information systems used in the industry. During a recession, the whole ICT department may be outsourced to a separate software service unit or to an IT house,

P. Tyrväinen
Department of Computer Science and IS, University of Jyväskylä, P.O. Box 35, 40014 Jyväskylä, Finland
e-mail: Pasi.Tyrvainen@jyu.fi

P. Tyrväinen and O. Mazhelis (eds.), *Vertical Software Industry Evolution*, Contributions to Management Science,
DOI 10.1007/978-3-7908-2352-3_4, © Springer-Verlag Berlin Heidelberg 2009

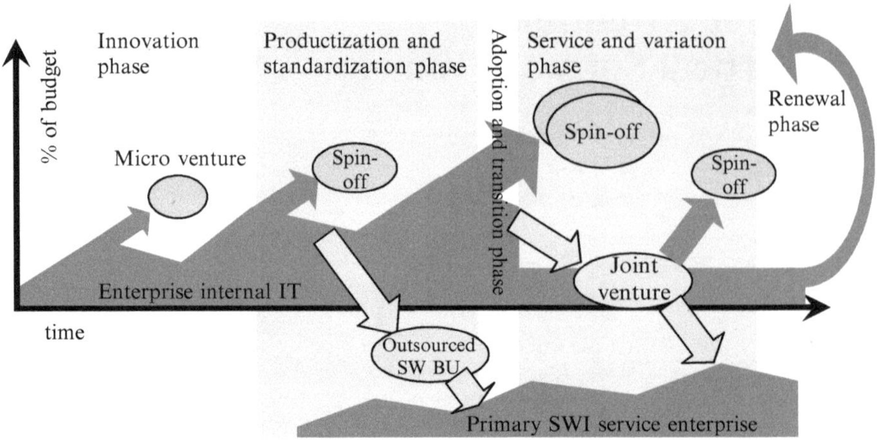

Fig. 1 Evolution of software development activity in a vertical industry enterprise. The horizontal axis represents time and the vertical axis total software spending

while in a growing economy this kind of unit may spin off as an independent venture to serve multiple firms. Typically, the vertical market leader maintains independent software development activity, while the competitors often join forces to standardize a common interface or platform, to specialize their activities and to provide jointly a broader offering than the market leader. These standardized interfaces also enable independent software vendors of the primary software industry to provide standard product offerings for a customer base with competitive pricing. The effort needed to implement a software product stemming from the evolution of a vertical market depends heavily on the degree of standardization of hardware and software interfaces (Garud and Kumaraswamy 1993), as well as on the standardization of the business processes the software systems support. This sets the minimum to the organization size sufficient for productizing the offering, and leaves the majority of the small software vendors out of the new market niche, due to major shortcomings in their competence portfolios. In this phase, one or some of the spin-off ventures or occasionally an existing primary software venture focusing on the emerging market are often able to create the first software products matching with the unifying functional requirements of the core processes and standardizing interfaces.

In the *Adoption and transition phase*, the micro ecologies formed around the standards grow fast and gain a higher user base, a higher vendor base and a higher market share. The vertical industry enterprises with internal software solutions start increasingly outsource their in-house software development (IT departments/business units), and adopt one of the competing standard software packages, which provide the whole product they need. If the number of the outsourced employees is in the range of 50 or more professionals, the enterprise may form a joint venture with a primary industry's software service provider. Typically this involves a few years' agreement on maintaining the existing core operational systems of the enterprise, or an assignment to help taking in use the emerging common software solutions.

In the *Service and variation phase*, one or two of the micro ecologies established in the former phases will have gained the position of a dominant design. As a result, the majority of new software development initiatives assume these products to be the ones to interface with. The software spending of the vertical industry enterprises at this point drops from the 5–15% level of the Productization phase to the level of 2% or less, as software is now considered a cost rather than a competitive advantage. A major portion of this spending is also shifted from in-house development to the purchase of software products and associated services. In addition, business process consulting and IT integration services are needed as external services. As one vertical may rather soon be saturated, the software enterprises seek to capitalize their knowledge also across vertical industries.

In the *Renewal phase*, some of the major vertical industry enterprises recognize a new business opportunity through software-supported business process reorganization and implement a new supporting software system with their in-house personnel, supported by external software implementation services. Once again, software is seen as a means to bring about a competitive advantage, and the evolution cycle starts another loop.

The advancement from one phase to another in the process of vertical software market evolution is driven by a variety of factors, which depend on the current phase of the evolution and of the overall market situation. Some of these factors are discussed next. The evolution may also stop to a phase due to lack of sufficient customer base or commonly accepted processes and technology standards needed for proceeding to the further phases.

2 Drivers of Change

Table 1 presents the drivers for change in a vertical software industry, along with the characteristic features related to the business strategies most applicable in each phase.

The vertical industry enterprises driving the adoption of a new technology seek to benefit from their internal (innovative) processes, in order to maintain their competitive advantage on the basis of process differentiation or improved cost performance in the innovation phase. Eventually, however, the competitors in the industry copy the best practices adopted in the industry, and, in the Productization phase they start acquiring software products from primary software industry ventures.

From the technology perspective, innovative enterprises that develop new technologies strive to secure beneficial technologies by filing patents and by using closed proprietary interfaces for software development. Leading hardware vendors are typically reluctant to share their knowledge with software companies and hence

Table 1 Changes in vertical industry processes and technology available set the pace for vertical software industry evolution

Drivers and features	Innovation phase	Productization phase	Transition phase	Service and variation phase
Process drivers (vertical industry)	Process innovation, competitive advantage (differentiation / costs), enterprise-specific processes	SW supported processes, process harmonization, repeatability between enterprises	Process standards adopted	Process optimization, IT process cost optimization
Software business	Customer-specific services, integrated HW/SW products, one application per enterprise	Productization, competing standards, micro ecologies, network development	Market share race, Competing ecologies	New development on the dominant SW platform, open source SaaS
Technology drivers	Technology innovation, competing with IPRs and closed interfaces, equipment vendors dominating SW market	Technology standardization within vertical market segment, multiple strong vendors	Fast technology adoption, market expansion	Technology dominance, bulk hardware with standard interface and large user base

create entry barriers for competitors: they implement software for operating their systems internally, thereby creating a unified hardware/software system.

The software business in this phase is limited to company internal or external software services. Independent software vendors cannot compete with the vendor who fully controls the technology platform needed for operating the software. Even if such software by independent software vendors appears, the proprietary interface can be changed, or the platform vendor may bundle free software with the hardware. As a result, unless the market is large and attractive enough, the software market evolution may stop in the innovation phase. Typical examples of software market stuck to the innovation phase include markets dominated by strong globally operating companies producing equipment such as elevators, weather sensors and game consoles.

Early console game vendors, for example, adopted this kind of strategy in the Innovation phase. They were competing primarily with new technology and platforms, while the process of playing remained unchanged. Game software applications were developed in-house. This kind of vertical integration was useful for optimizing the performance of the game consoles, which was critical for the customer preference. However, later on, criteria other than the platform became important when choosing between competing game offerings. The game markets demanded more effort on developing software applications and thereby forced this vertical industry to shift to the Productization phase. This vertical dis-integration has resulted in an increase in the number of game software developers in independent game companies, enabled improved products to be offered to the customers, and increased industry market volume. The increasing market volume also attracts more vendors, creates thereby more competition and enables also specialization. This further lowers the prices, thus attracting more customers, which again boosts the market.

As soon as the productized offerings meet the requirements of the processes of the vertical industry companies, the market enters the Transition phase. In this phase, typically one of the competing ecologies or standards reaches a dominating position, and its offering becomes the dominant design (see e.g., Murmann and Frenken 2006) in the Service and Variation phase. Some alternative designs may fit better to certain specialized purposes; a Macintosh computer platform long preferred by media processing industries is an example of that. Open source software development, which has already reached the Productization phase, is now making its presence felt in the Service and Variation phase.

The industry evolution may also stop in the Productization phase if the competing technologies and interfaces have insufficient critical mass to drive the positive cycle. There has been, for instance, a wide variety of fieldbuses used by various vertical industries to connect sensors and actuators with control units in real-time distributed control systems. Some of them have provided inexpensive solutions for industries with easier requirements, some of them have priced themselves out or are much less applicable to real needs. As a result, none of these fieldbus systems has been accepted widely as a dominant design. A long competition among Unix platforms is another example of the evolution stuck in the Productization phase.

3 Importance of Interfaces

The effort needed to implement a software system can be minimized by using existing software implemented by others. The use of such existing software commonly assumes a modularized system design, wherein the interacting modules are integrated through interfaces. These existing subsystems act as the containers for knowledge transfer from external organizations to the software system developer. Thus, the availability of system components with easy-to-use interfaces reduces greatly the knowledge required by the developers. If the interface is standardized, the software developer does not have to understand the details of the system used, whereas employing nonstandard interfaces with variations often demands a profound knowledge of the system. The requirement to master a large number of proprietary technologies and to have the necessary industry process knowledge can be too much for most of the software development organizations. This may limit the critical mass of the organization needed for reaching the next phase in the vertical industry evolution.

Figure 2 demonstrates the appearance of new interfaces and software layers along the evolution. In the innovation phase each of the hardware vendors (A, B and C on the left) has implemented proprietary software interfaces to their hardware

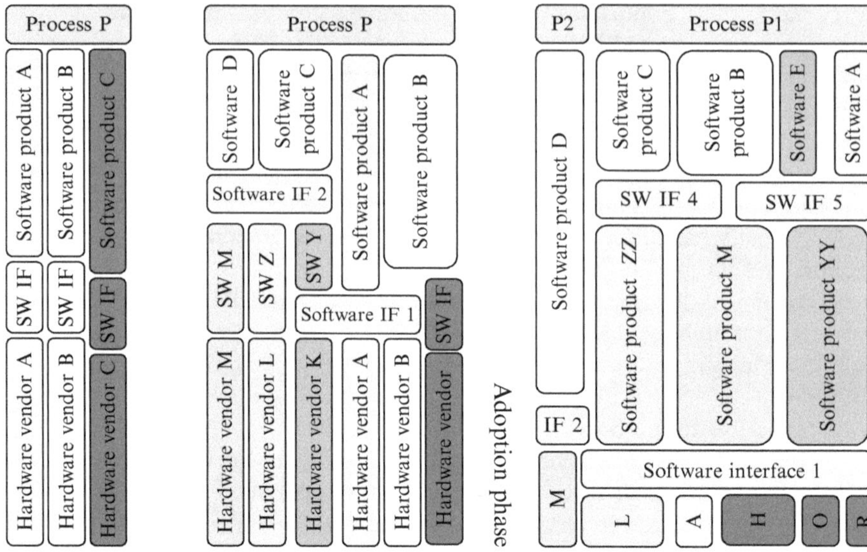

Fig. 2 Software interfaces and vertical dis-integration along vertical software industry evolution

(labelled "SW IF"). Each of them has also developed the application (labeled "Software product" A, B and C respectively) for supporting user process ("Process P"). This example could represent three elevator vendors or early market of game console vendors, each developing the game software products for their customers. In case vendor C is the market leader with leading proprietary technology and the best set of patents, vendors A and B may join their forces and define a common interface ("Software IF 1" in the middle). This would drive the evolution into the Productization and standardization phase (or Productization phase, in short). In this case the application offering provided on top of Software Interface 1 would be competitive and attractive as it would include not only the products of vendors A and B, but in addition software offerings of independent software vendors (ISVs) building their applications on top of Software Interface 1 (or interfaces built on top of it). Increasing number of customers would attract more ISVs and hardware vendors (vendor K) to support the interface, providing a growth cycle for the micro ecology round Software Interface 1.

In the Productization phase the competing microecologies try to find the optimal structure to gain maximal user acceptance within the industry. The competing interfaces may also build on top of each other, the way S60 platform is building on top of Symbian in the domain of operating systems for mobile phones while competing with Android and Microsoft Windows Mobile. In case one or more of the interfaces are successful in meeting the essential requirements of the majority of the application domains they will enter the Adoption phase. Most of the applications start to use these interfaces as a platform due to their large potential customer base. After a race for market share, one of the interfaces is likely to reach the position of a dominant design ("Software interface 1" on the right) in the Service and Variation phase. There may still exist some niche markets for competing interfaces ("IF 2") such as Apple competing with Microsoft in the operating system interfaces. New software vendors (ZZ, M, and YY in Fig. 2) will then consider it useless to compete with this dominant design and start, instead, building additional software layers on top of it. For example, after the SQL interface became the dominant design for accessing databases, further layers , such as ODBC, JDBC etc., were build on top of it.

Figure 2 represents the reasons for vertical dis-integration that is realized in increasing number of software layers provided by independent software vendors. This development will also reduce the effort needed for developing the software needed to support a specific process or to interface a specific hardware, as there exists an increasing volume of software that can be used as a base-line for any new software.

However, new software requires typically multiple interfaces to interface with multiple hardware-based functionalities. As an example, a software organization needed for implementing a basic retail banking system is considered in Table 2. This system needs to interface with the customers, be able to deliver notes and communicate with a teller, and have a mainframe computer with a database management system connected to the server (on the left). For a long time, the interfaces available to these technologies were proprietary, thereby limiting the number of vendors to a few (see configuration A in the middle). Once the technologies evolve, the interfaces needed are standardized, hence making it easy for a large number of companies to implement the system.

From the viewpoint of the interfaces, large IT service companies with a well equipped competence pool tend to dominate Innovation Phase markets. Smaller product companies appear more often in the Productization phase where emerging standards can be utilized. In the last phase, the dominating design again favors large companies. Note also, that a small company adopting configuration B in the example (which is based on common internet technologies) has dropped the note delivery interface from the system offering. This means that this inexpensive configuration can be used by retail banks which are willing to drop this functionality

Table 2 An example: Alternative interfaces used by a retail banking application. The interfaces used in configuration A in the middle are mainly from the Innovation phase and require a high level of competence (available only in a few software enterprises or in-house). The interfaces in configuration B on the right are from a later date and only require knowledge that is available with millions of software developers

Hardware interfaces needed for automation of the process	A: Traditional configuration in the innovation phase	B: Configuration using only standardized interfaces
Customer user interface	ATM user interface (I)	Web browser(S)
Note delivery interface	ATM messaging IF (I)	– (excluded)
Teller interface	Windows PC (S)	Windows PC (S)
Server hardware	Mainframe (I)	Linux/Windows (S)
Database	Mainframe database(I)	SQL database (S)

from their offering or are able to outsource it. Thus, the development of the set of technologies needed by a vertical industry (retail banking) impacts the cost/functionality of the software offering available to the vertical industry companies. But the lack or emergence of a new software offering can also impact the business model of the industry by creating a market for ATM outsourcing.

4 Shift of Software Development Activity

During the market evolution, the focus of a vertical market shifts inevitably from the use of internal resources to an extensive use of software products and professional software services provided by the primary software industry. In Fig. 3, this is shown by dividing the IT spending on software related activities to internal software development, external software services, and software products. The form of the increasing share of the spending going to the primary software industry resembles a horn and thus this model is also called as the Trumpet Model.

In the Innovation phase, a major part of the activity takes place within the enterprises. As the market develops, the share of software products and services increases. Some verticals may remain in the early phases for years (such as several manufacturing segments with insufficient volume for software productization), while most of the verticals proceed further. This development is visible also in the total figures of software spending. According to EITO (EITO 2004), 52–59% of

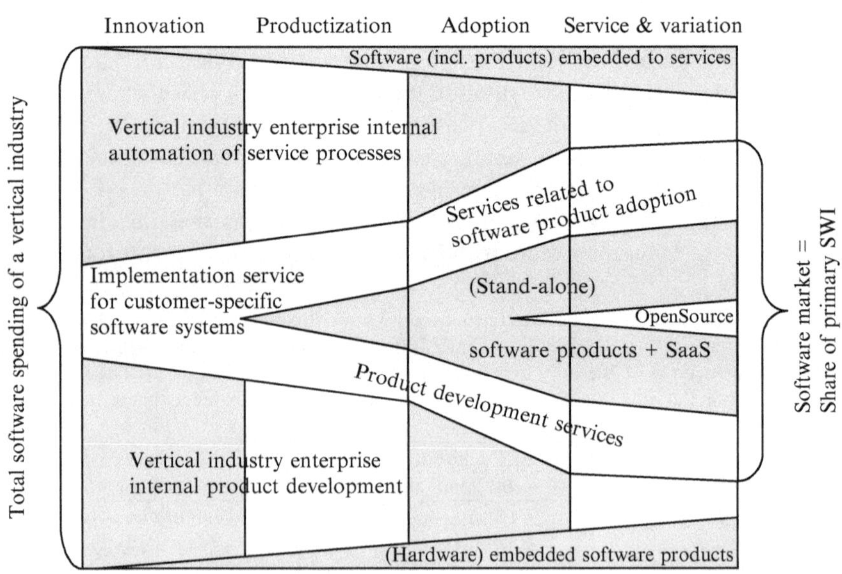

Fig. 3 The shift of vertical industry software spending from in-house development to use of external products and services

IT spending in Europe, Japan and USA went to internal software development in 1993, while purchased services and software packages took 14–32% and 10–29% of the spending, respectively. Ten years later these three segments were roughly equal indicating a major drop of internal spending in respect to purchased services and software products (Tyrväinen et al. 2004, 2008). There are still industries where large vertical industry enterprises want to maintain a strong control over innovative software-supported services and keep a major part of their software development in-house. This especially applies to industries where technology or process innovations play an important role, such as the car industry, some other manufacturing and electronic industries as well as opportunity seeking service industries.

In terms of visible revenues, the secondary software industry is considerably worse than the primary software industry, as the added value of the software is sold along with a hardware product (e.g., a mobile phone) or included in the price of a service (e.g., banking service fees).

The evolution may also be halted in the early phases if the market size is insufficient to attract enough competition to the segment, or if the dominating companies are unwilling to proceed. Especially in specific market niches of manufacturing industries, some of the dominating players are unwilling to open their software interfaces for open competition.

Overview of Telecom Operator Software Market

Eetu Luoma, Lauri Frank, and Mirja Pulkkinen

Telecommunications is an essential enabler of modern societies and a global vertical industry, providing communication and information services, with its annual revenue of over trillion euros. In this book, a company providing these services is referred to as a *telecom operator* or *communications service provider* (CSP). CSPs create value by offering connectivity to and via an infrastructure of networks for transferring signals. The core business processes in a telecom operator company therefore revolve, on one hand, around physical networks and their maintenance, and on the other, managing their customers and their use of the connectivity services over the networks.

In the past, CSPs' operations focused more on the development and management of network infrastructure. Eventually, as the need for standardization of the systems and operations became apparent, International Telecommunication Union (ITU) provided a set of specifications covering different functional management areas. More recently, the telecom business realized the importance of adopting the so-called "lean" approach, and the focus of standardization efforts switched to streamlining the business processes (Thomsen 2006). In particular, TM Forum provides guidance for such work through a set of reference architecture specifications. To provide an insight into the types of telecommunication software systems and into the vertical software market segmentation, a brief overview of the activities by ITU and TMForum is given below.

E. Luoma (✉)
Department of Computer Science and IS, University of Jyväskylä, P.O. Box 35, 40014 Jyväskylä, Finland
e-mail: Eetu.Luoma@jyu.fi

P. Tyrväinen and O. Mazhelis (eds.), *Vertical Software Industry Evolution*, Contributions to Management Science,
DOI 10.1007/978-3-7908-2352-3_5,

1 ITU

The ITU is the United Nations agency for information and communication technologies. In the domain of telecommunications management, their central concept, denoting management activities of CSP, is the telecommunications management network (TMN). It is an architecture encompassing processes and atomic tasks for planning, provisioning, installation, maintenance, operation and administration of telecommunications equipment, networks and services. The architecture is divided into layers including business, customer, service, network and element management layers (ITU-T 2000). The telecommunications management is therefore concerned with operational, that is, day-to-day processes for service fulfilment, service assurance and billing as well as infrastructure development processes.

Operations support systems (OSS), business support systems (BSS), network management systems, and element management systems corresponding to the TMN layers constitute the CSPs system environment. The main subject of this book is the software systems supporting the processes specific to this vertical industry, namely the OSS and BSS. The term *Operations Support System* (*OSS*) refers to software systems supporting network management processes such as maintaining network inventory, provisioning services, configuring network components, collecting and mediating usage information and managing faults. The term *Business Support Systems* (*BSS*) typically refers to software systems supporting customer management processes including taking orders, providing customer service, processing bills, and collecting payments (Terplan 2001).

2 TM Forum

The TM Forum (also TeleManagement Forum, TMF) is a global nonprofit consortium that develops business and technical frameworks, specifications and guidelines for designing telecommunication, media and content-based services. The Next Generation OSS (NGOSS) initiative organized through TMF supports the ongoing business process automation work of the CSPs, software vendors and systems integrator, through the provisioning of process maps, data models, and applications maps. Therefore, NGOSS can also be considered as the reference enterprise architecture for the industry and as an intended standard in defining and developing the OSS/BSS components. The initiative has produced:

- The enhanced Telecom Operations Map (eTOM) defining CSPs processes
- The Shared Information and Data (SID) framework supporting harmonized information definitions to be used as a common language in NGOSS-based applications
- The Telecom Application Map (TAM) defining the functionality of telecom supporting applications or services

- Technology Neutral Architecture (TNA) defining concepts for managing the lifecycle of the OSS/BSS and the requirements for and services of an NGOSS-compliant architecture

For the purpose of analysing the OSS/BSS market, the eTOM and the TAM above provide a good starting point in explaining how CSPs conduct their business, and therefore, which software systems they deploy in order to support their operations. The eTOM framework describes a standard structure, terminology and classification scheme for all the CSP processes at different levels of detail, and according to their priority for the CSP's business. The framework thus provides a basis for restructuring CSP operations and also for general working agreements with software providers. The eTOM framework is represented as a process model in a series of groupings using hierarchical decomposition (Kelly 2003).

The three major process areas described by TMF (2005) are:

1. *The Strategy, Infrastructure and Product process area* – includes processes needed in order to develop CSP strategy, build infrastructure, develop and manage products, and to develop and manage partnerships and the supply chain.
2. *The Operations process area* – includes vertical end-to-end processes for service fulfilment, service assurance and billing processes. These processes are sometimes referred to as customer operations processes and include both CSPs day-to-day processes and operations support and readiness processes. The latter are responsible for providing management, logistics and administrative support for customer operations.
3. *The Enterprise Management process area* – includes basic business processes required to run any business. These processes focus on enterprise level, and include, e.g., financial management processes, human resources management processes, etc.

In the operations process area, the eTOM framework depicts the three vertical operations as end-to-end processes of the telecommunication industry. The vertical process group of *fulfilment* is responsible for providing customers with their requested services. The fulfilment workflow advances from selling and order handling through service configuration and activation to resource provisioning. The processes in the service *assurance* group are responsible for the execution of proactive and reactive maintenance activities to ensure that services provided to customers are available at the agreed SLA or QoS performance levels. The processes are about collecting data on problems and service quality parameters at both customer and network interfaces as well as aggregating and analyzing the data. The *billing* process group is responsible for the production of timely and accurate bills, resolving billing problems, processing customer payments, and performing payment collections. The tasks in this process group basically involve collecting data records on usage of the services, aggregating and rating service usage and charging customers according to the usage.

The eTOM operations process area also includes the horizontal layers of customer management, service management and resource management that partly correspond to the TMN layers. The *customer relationship management* processes are responsible for having knowledge of customers' needs and the acquisition, enhancement and retention of a relationship with a customer. The *service management* processes are responsible for knowledge support in the areas of communication, media and information services and the management and operations of these services for customers. The *resource management* processes are responsible for maintaining knowledge of resources (application, computing and network infrastructure), as well as for managing all the resources to support services for customers.

The Telecom Application Map (TAM) defines the properties of OSS/BSS by combining specifications in the eTOM and SID frameworks into functioning applications. TAM is a framework for both those who sell and buy telecom software. TAM is in line with the vertical and horizontal grouping of the eTOM framework and basically it perpetrates the view that there might be at least one application for every single vertical process.

3 OSS/BSS Market

3.1 Market Segments

In order to examine the vertical software market, an interpretation of the market in the form of segmentation is needed. Several market research companies (e.g., Dittberner, Lightreading, OSS Observer) have followed the evolution of the telecom software market and classified firms operating in the industry into various market segments. The segmentation of the OSS/BSS markets followed in this book is based (1) on the definitions provided by these three market research companies, (2) on the Telecom Applications map, as well as (3) on the defined properties of the OSS/BSS software. Here, the OSS/BSS market has been divided into 13 segments, which have been positioned on top of the above-mentioned vertical processes and horizontal layers. Figure 1 presents this mapping of the OSS market segmentation on top of the operations process categories of the eTOM framework.

3.2 OSS/BSS Market Volume

In order to study the telecom software market, we use commercial data collected by Dittberner Associates, Inc. (for the description of the data used, the reader is referred to Section "Data Sources Used"). The total volume of the OSS/BSS market in 2006 was nearly 30 billion USD. Out of the total number of companies in the market, one-third are ICT service providers that produce 55% of the total revenues. Overall, the OSS/BSS vendors received 64% of their revenue from services and

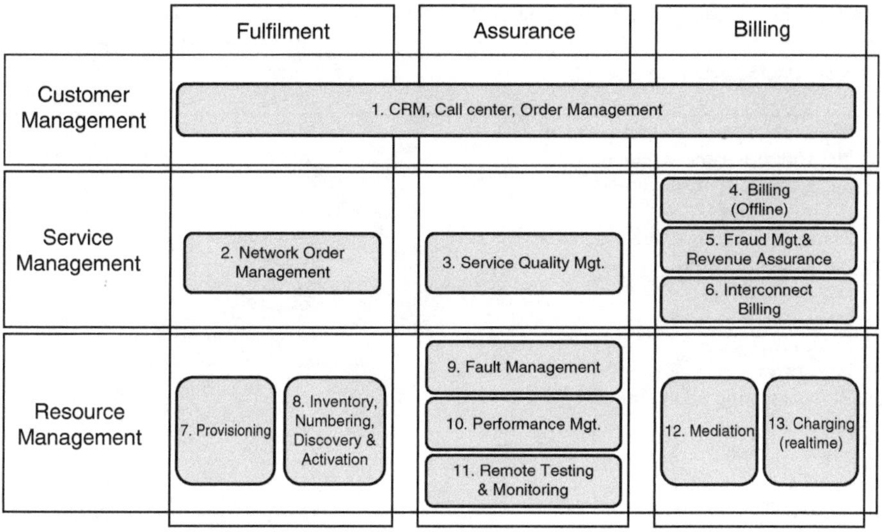

Fig. 1 OSS/BSS market segments on top of the eTOM operations process group

36% from software products. The segments under the vertical process group billing have the biggest volume of the software product segments with close to 6.3 billion USD, followed by the segments under assurance (3.7 billion USD), fulfilment (1.5 billion USD) and the customer relationship management (0.85 billion USD) spanning horizontally at the customer interface shown in Fig. 1. The segment-specific total revenues are listed in Table 1 below.

In 2006, the largest geographical market for OSS/BSS was in North America (46% of the total market), although the market had not grown there at all compared to year 2000. In contrast, over this 6-year-period the OSS/BSS market had grown in EMEA over 40% (35% of total), in Asia Pacific over 64% (14% of total), and in Latin America over 93% (5% of total).

3.3 OSS/BSS Market Scope Strategies

The competition in the OSS/BSS market can be observed through strategies adopted by a software vendor with respect to its market scope. Based on Fig. 1, it can be hypothesized that an OSS/BSS vendor might consider four distinct market scope strategies, ranging from narrow to broad:

1. *Niche strategy* (also "narrow strategy"). Here, the vendor provides software only for one market segment presented in Fig. 1, except for the CRM, which ranges through several layers. The companies deploying a niche strategy possess distinct knowledge or strong customer relationships on one or several segments.

Table 1 The number of companies and total volume of software product segments and services for the year 2006

Software product segments	Number of companies	Total volume (million USD)
1. CRM, call center, order management	28	$852.3
2. Network order management	18	$168.3
3. Service quality management	27	$363.8
4. Billing (offline)	33	$3219.1
5. Fraud Mgt. and revenue assurance	25	$393.7
6. Interconnect billing	28	$292.8
7. Provisioning	30	$672.2
8. Inventory, numbering, discovery and activation	22	$359.9
9. Fault management	25	$563.0
10. Performance management	29	$1156.1
11. Remote testing and monitoring	16	$503.3
12. Mediation	42	$453.9
13. Charging (realtime)	33	$1929.7
Software products total	131	$10928.1
Services total	65	$19038.7
OSS/BSS market total	196	$29966.8

Market entry using a niche strategy usually takes place simultaneously with the introduction of a new technology, e.g., a new network interface or a middleware.

2. *Vertical strategy*. The vendor provides software in at least two market segments, but only in one of the vertical process groupings presented in Fig. 1. To be able to provide software for managing vertical processes, a vendor is required to have special knowledge on both the particular processes and the network interfaces related to this vertical process. This kind of a strategy is typical for the established firms in the OSS/BSS market.
3. *Layer strategy*. Software for multiple market segments (or the single segment of CRM) is provided within the same horizontal layer presented in Fig. 1. This strategy refers to focusing on supplying software on one of the horizontal layers (customer, service or network management). Established companies are likely to prefer strategies close to the network management layer at the bottom, whereas new entrants more likely appear close to the customer interface on the top. On the one hand, this is due to the similarity of CRM processes across industries and, on the other, due to required domain expertise on network technologies that cannot be transferred from other vertical software markets.
4. *Broad strategy*. A vendor possessing broad expertise on both telecom specific technologies and processes. Such a vendor provides OSS products for several segments, at least one of the vertical segments and/or one of the horizontal layers as described above and in Fig. 1.

The market scope strategy used by the vendors has yet another dimension: the extent of vendor's focus on telecommunications. This is related to the evolution of software industry, where software production generally shifts gradually from in-house production, first to external producers, and then further to general, i.e., horizontal software firms (see Section "Shift of Software Development Activity" in Chapter "Model for Evolution of a Vertical Software Industry"). Vendors with a telecommunications industry focus can be divided into three categories: (1) telecommunications-specific vendor with more than 90% of revenue from the telecommunications industry; (2) telecommunications industry dependent vendor with more than half (50–90%) of its revenue from the telecommunications industry; and (3) horizontal software vendor with a majority of revenue from the outside of the telecommunications industry (i.e., <50% of revenue from the telecommunications industry).

Together, the breadth of a firm's operations and the extent of focus on the telecommunications industry can be used in order to classify the OSS/BSS vendor strategies: the breadth of a firm's operations can be used as one dimension for measuring an OSS vendors' market strategy, while the second dimension can be derived from the extent of focus on the telecommunications industry. Figure 2 presents the classification of the types of OSS/BSS vendor strategies.

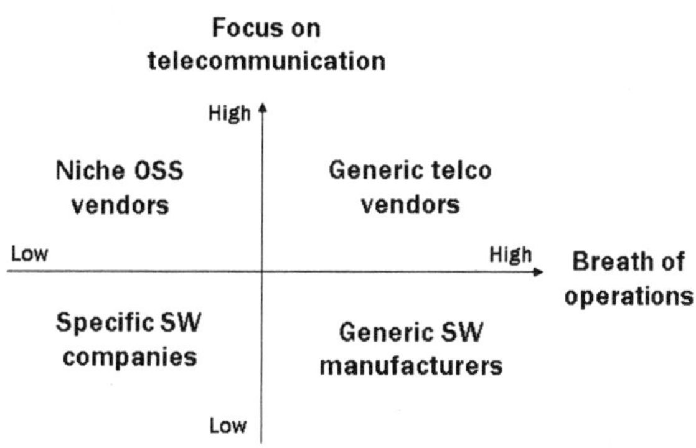

Fig. 2 A classification for OSS/BSS vendor market scope analysis

Table 2 Number of vendors and average revenue by breadth of operations and telecommunications focus in the year 2005

	Telecommunications focus (% of revenue from telecommunications industry)							
	<50%		50–90%		>90%		Total	
Breadth of operations	*N*	Average OSS revenue	*N*	Average OSS revenue	*N*	Average OSS revenue	*N*	Average OSS revenue
Narrow-1	8	M$ 14.9	5	M$ 20.7	34	M$ 19.9	47	M$ 19.1
Vertical-2	5	M$ 155.0	13	M$ 66.4	22	M$ 40.6	40	M$ 63.3
Layer-3	3	M$ 124.2	1	M$ 80	4	M$ 29.4	8	M$ 62.3
Broad-4	4	M$ 82.3	8	M$ 272.7	18	M$ 224.6	30	M$ 218.4
Service providers	43	M$ 217.5	8	M$ 39.4	12	M$ 103.8	61	M$ 171.5
Total	63	M$ 593.9	35	M$ 407.2	90	M$ 418.3	186	M$ 534.6

In order to test the above classification of market scope strategies, Frank et al. (2007) analyzed OSS vendor revenues from 2005, using the data covering a total of 186 companies operating in the OSS/BSS market. The service providers, such as systems integrators, were excluded from the analysis and placed into a separate category outside the strategies listed. Then, the software product vendors were categorized according to the breadth of their operations and focus on the telecommunications industry. Table 2 presents the number of vendors in these categories.

Based on this analysis, a conclusion can be made that all the hypothesized strategies were deployed in 2005: the OSS market has vendors with narrow niche, vertical, layer and broad strategies, and companies classified as service providers (with no software products). About half of the vendors operating in the OSS markets can be considered to focus on the telecommunications industry. About a third of the firms operating in these markets receive less than a half of their revenues from telecommunications.

Part II
Current State of the Telecom Software Market

Communication service providers have been active in adopting software technology for OSS/BSS processes as a means to improve productivity. The first section of this chapter starts with a general analysis on when and how software can be used as a means to improve enterprise productivity, and the second section continues with a small-scale quantitative analysis on the impact of software usage to CSP performance. The third section uses interviews of CSP representatives to gain internal view to the use of software in telecom operators. The interviews from Europe and the Far East disclose CSPs' criteria about when to buy software instead of making it in-house and give some background on the CSP business environment in China and Taiwan. Section four targets the OSS/BSS software market evolution and analyses it using multiple market development models.

Process Improvement Through Software

Olli Martikainen

1 Investments in IT Improve Productivity when Connected to Organizational Changes

The productivity increases resulting from ICT in the industry and in society are of a rather recent origin as shown by Mika Maliranta and Petri Rouvinen from ETLA (2003). According to their research it seems that the additional productivity achieved typically ranges from 8 to 18%. The effect on services tends to be larger than on manufacturing. The effect is often manifold in younger and can even be negative in older firms. Since organizational changes are easier to implement in younger firms and recently established firms have by definition a new structure, this can be interpreted as evidence for the need for complementary organizational changes. Manufacturing firms seem to benefit from ICT-induced efficiency in internal communication whereas service firms benefit from efficiency in their external communications.

Also the results of Maliranta and Rouvinen provide direct and indirect evidence on the importance of competition, education, innovation, organizational change and entrepreneurial dynamics in the adoption and efficient use of ICT. Possible lags in the effects of ICT are not very well known. Furthermore, timing between the implementation of ICT and complementing organizational changes remains an open question. According to the study, one would expect that the two are implemented simultaneously, but anecdotal evidence would seem to suggest that organizational changes that follow show a considerable time lag.

In the literature, only a limited amount of knowledge has so far been made available about what might be the best practices in organizing ICT-assisted work.

O. Martikainen
Department of Information Processing Science, University of Oulu, P.O. Box 3000, 90014 Oulu, Finland
The Research Institute of the Finnish Economy (ETLA), Lönnrotinkatu 4 B, 00120 Helsinki, Finland
e-mail: olli.martikainen@etla.fi

P. Tyrväinen and O. Mazhelis (eds.), *Vertical Software Industry Evolution*, Contributions to Management Science,
DOI 10.1007/978-3-7908-2352-3_6,

It is nevertheless obvious that for vast majority of workers, job descriptions will ultimately differ considerably from the current ones. This in turn will have consequences in the labor market and for society at large, which should be studied.

In their further research Maliranta and Rouvinen (2004) study productivity effects of portability as well as wired and wireless connectivity using company level data of Finnish ICT. They found that a computer with only processing and storage capabilities boosted labour productivity by 9% (corresponding to 5% output elasticity), portability by 32%, wired connectivity by 14%, and wireless connectivity by 6%. These findings are in line with previous data, and comparisons to ICT costs suggest that firms equate marginal costs and returns. While increases in ICT employment can no longer be a major source of productivity growth in developed economies, the studied new characteristics can.

2 Technological Innovations Need Application Innovations to Improve Firm Performance in the Software Area

Technical progress and innovation have long been recognized to have the central role in economic growth (e.g., Kuznets 1930; Schumpeter 1939). However, on firm level the effect of innovation to firm growth has not always been found. The empirical research of 21 studies referred to in Del Monte and Papagni (2003) gives the following result: research intensity measured by the R&D to sales ratio indicates a positive effect on growth rate of firms in four out of seven studies.

These same authors analyzed also the relation between innovation and performance of 884 Italian firms during 1992–1997. Their results confirmed that the growth rates of sales and personnel in firms conducting R&D were larger (56% and 18%, respectively) than those of firms without R&D (47.7% and 10.4%, respectively) with a statistically significant difference (see Del Monte and Papagni 2003).

A later study of Japanese manufacturing firms (Yasuda 2005) shows that R&D expenditure per employee has a significant positive effect on firm growth. A similar study of Taiwanese electronics firms (Yang and Huang 2005) shows that an increase in R&D induces a higher growth rate, and that this impact is larger in small firms.

Firms' innovation intensity has usually been measured using their R&D expenditure or patenting intensity as variables. In later studies also other measures have been included, such as comparison of product and process innovations (Harrison et al. 2008) or comparison of product innovation, process innovation, organizational changes and marketing innovation (Morone and Testa 2008). In the first case, firm level data from France, Germany, Spain and UK indicates that, in manufacturing, product innovation is associated with employment growth whereas process innovation tends to displace employment. In the second case, the results from 2,600 Italian manufacturing firms indicate that process innovation and organizational change have a stronger effect on a firm's growth than product innovation and marketing innovation.

Table 1 Innovations and firm growth: findings available in literature

Authors	Data	Industry	Main results	Method
Kuznets	1930		Product innovations are central for economic growth	History of technology
Schumpeter	1939		Historical role of technological innovation	History of technology
Del Monte, Papagni	1992–1997	Italian firms	The growth rates of sales and personnel in firms conducting R&D were larger than those of firms without R&D	Statistical
Yasuda	1992–1998	Japanese manufacturing	R&D expenditure per employee has a significant positive effect on firm growth	Statistical
Yang, Huang	1986–1999	Taiwanese electronics	Increase in R&D induces a higher growth rate	Statistical
Harrison et al.	1998–2000	French, German, Spanish and UK firms	In manufacturing, product innovation is associated with employment growth whereas process innovation tends to displace employment	Statistical
Morone, Testa	2004	Italian manufacturing	Process innovation and organizational change have a stronger effect on a firm's growth than product innovation and marketing innovation	Statistical

The results explained above are summarized in Table 1.

In a recent study by Jyrki Ali-Yrkkö and Olli Martikainen (2008) on innovations in Finnish software industry, innovations were divided into two groups: technological and nontechnological innovations. The firms were asked to describe the impacts of different types of innovations on firm growth (2008). The survey was conducted in spring 2008 as a joint effort by Helsinki University of Technology (Software Business Lab) and University of Turku (Software Product Development Research Group).

To measure different dimensions of innovations, the Innovation radar approach by Sawhney et al. (2006) was applied. The innovation radar is composed of 12 dimensions of innovation:

1. Offerings
2. Platforms
3. Solutions
4. Customers
5. Customer experience
6. Value capture

7. Processes
8. Organization
9. Supply chain
10. Presence
11. Networking
12. Brand

Out of these 12 dimensions, three (offerings, platform and processes) are clearly related to technology development. Offerings mean products and services, and innovation of these requires the creation of new products of services. A platform is a set of common components, assembly methods or technologies that serve as building blocks or modules for a portfolio of products and services. Processes are a configuration of activities used to conduct internal operations. Innovations along offerings, platform and processes require very often R&D. The other nine dimensions on the innovation radar are nontechnological innovations.

The study analyzed the impact of innovations on firm growth using data on Finnish firms operating in the software industry. The dataset of 267 observations consisted of firms that are mostly small and private.

The results based on the data suggest that innovations indeed have a statistically significant positive relationship with firm growth in the software industry. However, this positive relationship could be found only in the cases where firms had accomplished both technological and nontechnological innovations. Hence, pure technological innovations alone or R&D activities alone did not seem to increase firm growth.

Effect of Software on CSP Performance

Oleksiy Mazhelis, Pasi Tyrväinen, and Kimmo Suojapelto

In order to verify how (if at all) CSP's investments in software affect their business performance, the relationship between (i) CSP's spending on software and (ii) its key performance indicators (KPIs) is considered below. The investments in software are represented by internal, external, and total software spending a year or two years prior to the KPI values, while the KPIs include average revenue per user (ARPU), revenue, net income, EBIDTA, and change in revenue (as an indicator of company growth). Besides, Opex and Capex along with R&D expenses are used as control variables. We use the data collected for the years 2004–2007 (revenue, net income, EBITDA, market capitalization, R&D, CAPEX, OPEX), and 2004–2006 (ARPU). The data is used to construct regression models wherein the dependencies between KPIs and software spending are evaluated in terms of the variance explained by each of the independent (or control) variables.

The dependency of CSP's KPIs on software spending, and Opex, Capex, and R&D expenses are summarized in Fig. 1. The thickness of arrow lines indicates the magnitude of the variance explained (thick line indicates the coefficient of determination of $R^2>0.5$, thin line – $0.1<R^2<0.5$, and dotted line – $0.05<R^2<0.1$. The type of the arrows designates the sign of the regression coefficient: the double (single) line corresponds to the positive (negative) value of the coefficient.

As suggested by the figure, the CSP's spending on internal software development may cause a reduction in the revenue. This can be explained by the fact that, being in the productization phase, the OSS/BSS software brings relatively little competitive advantage to the CSPs (since competing CSPs can be assumed to deploy similar software); and therefore allocating extensive resources to in-house OSS/BSS software is unlikely to result in a high return on investment, as compared with alternative investing opportunities.

O. Mazhelis (✉)
Information Technology Research Institute, University of Jyväskylä, P.O. Box 35, 40014 Jyväskylä, Finland
e-mail: Oleksiy.Mazhelis@jyu.fi

P. Tyrväinen and O. Mazhelis (eds.), *Vertical Software Industry Evolution*, Contributions to Management Science,
DOI 10.1007/978-3-7908-2352-3_7,

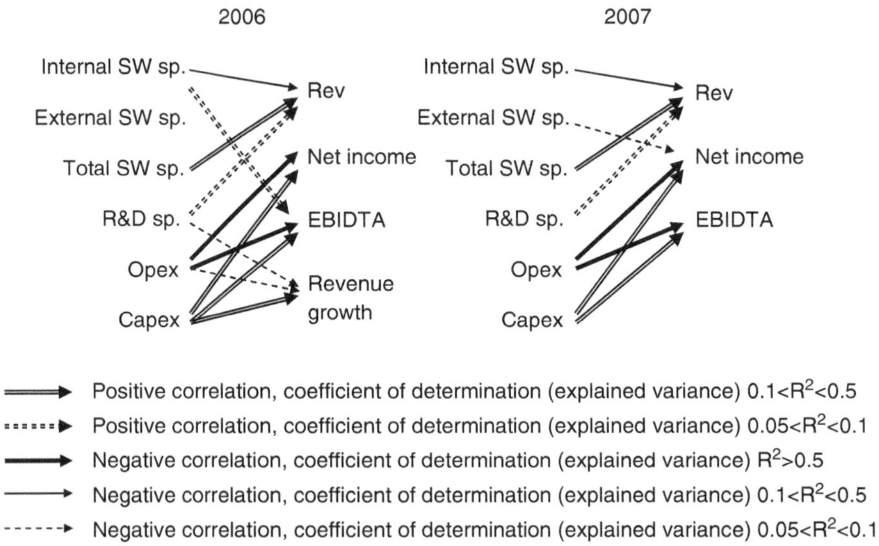

Fig. 1 Dependency between KPIs of a CSP (ARPU, revenue, net income, EBIDTA, and change in revenue) and software spending, Opex/Capex, and R&D spending

One such alternative investing opportunity, which is apparently more cost-efficient, is acquiring software from independent software vendors: as suggested by the analysis, spending on external software acquisitions is positively correlated with the CSP's revenue.

As expected, the data indicates the negative effect of Opex and the positive effect of Capex on the net income, EBIDTA, and the revenue growth. The R&D spending, according to the data, has a slight positive effect on the revenue.

Stepwise linear regression models were constructed for analyzing the effect of software spending (total, internal and external software spending, R&D, Opex, Capex and other variables) on ARPU and other KPIs. One of the most interesting models generated is presented in Fig. 2. In this model, ARPU 2006 was explained with a single independent variable, in terms of software spending of 2005. As can be seen, this single variable explains 51% of variance, with a positive coefficient of 0.71 (cf. Fig. 2).

It should ne noted that the value of ARPU may be largely affected by the average level of wealth in the country, since a low average level of income in the country limits the possibilities for the growth in the customers' spending on communications, and therefore limits ARPU. Indeed, according to the data, ARPU, besides being correlated to software spending, is also correlated with the Gross Domestic Product (GDP) at Purchasing Power Parity (PPP) per capita in the country (see the values of the Spearman's rho coefficients in Table 1).

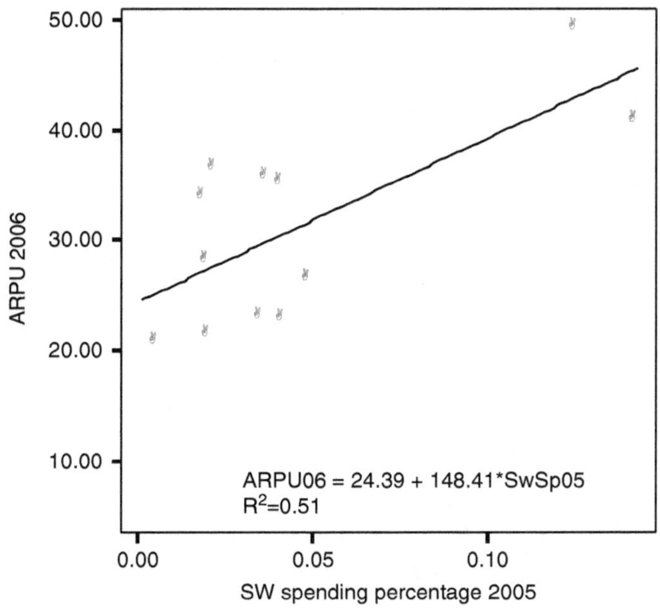

Fig. 2 ARPU 2006 explained in terms of software spending during 2005 (note: the estimation of regression coefficients is based on only 12 cases)

Table 1 Correlation (Spearman's rho) between ARPU, software spending, and GDP PPP (the-triple in each cell includes the Spearman's rho coefficient, the significance, and the degree of freedom). Correlation coefficients significant at the 0.05 level are shown with *

	GDP PPP	CSP's software spending 2005	CSP's software spending 2006
ARPU 2005	0,78*	0,55	0,70*
	0,00	0,06	0,01
	27,00	12,00	12,00
ARPU 2006	0,81*	0,53	0,69*
	0,00	0,08	0,01
	27,00	12,00	12,00
GDP PPP		0,31	0,16
		0,18	0,44
		20,00	24,00

The limited availability of KPI and software spending data for CSPs restricted the target set of CSPs in this analysis. However, reaching statistically significant correlations with a data set of this size encourages further empirical analysis on the impact of SW usage making use of industry-specific performance indicators.

Insights from Operator Interviews

Mirja Pulkkinen and Jari Veijalainen

The previous section analysed use of software to improve productivity from a general perspective and provided a quick statistical analysis of software usage to the performance of a CSP. Before conducting elaborated statistical analysis on CSP software usage, we need to gain some insight into operator business and understanding on how operators see the role of software in their business as well as how they acquire software.

Telecommunications operators run a software intensive business. Currently, there exist a variety of access networks which form the technological infrastructure for providing communications services. Each network technology may require specific software applications for running services on that network. As seen in the previous sections, the OSS/BSS software is composed of numerous applications each for a rather narrow area of functionality. Further, in the operator business, growth seems to come in the developed countries from services built on top of plain network connectivity. The CSPs provide access to the Internet and to diversifying mobile services (besides telephony, also data transfer, email and chat services, and as a growing area also media, including video and TV), as well as a growing number of other value added content services. PSTN telephony is commonly facing phase-out in the portfolio of CSP services. All this adds to the number of applications a CSP has to maintain in their application portfolios.

As a means of exploring the telecommunications software in general and exploring the reasons behind the market figures in particular, interviews with representatives of CSPs were conducted. North and Central European operators were interviewed first; these interviews served as a background for the other explorations and informed on the situation on the old continent. In a second set of interviews, Asian operators helped to complete the picture. The interviews and the insights obtained from them are described next.

J. Veijalainen (✉)
Information Technology Research Institute, University of Jyväskylä, P.O. Box 35, 40014 Jyväskylä, Finland
e-mail: mirja.pulkkinen@titu.jyu.fi

P. Tyrväinen and O. Mazhelis (eds.), *Vertical Software Industry Evolution*, Contributions to Management Science,
DOI 10.1007/978-3-7908-2352-3_8,

1 Interviews in Europe and Asia

1.1 Interview Process

The scheme to interview was semi-structured (Fontana and Frey 2000), i.e., topical areas were given and the interviewees had the opportunity to direct the conversation, ask back and reflect on the topic. In this way, they had the opportunity to bring into the discussion the points they considered as important regardless of whether they had been asked questions about them. However, to guide the conversations, a list of topical areas was shared with the interviewees. That ensured that the resulting information was useful for comparing the telecom industry software developments to the general model of the software market development. The list of topics included, among other things:

1. Own software development vs acquiring software products or services.
2. Procurement of applications and systems, selection of providers of software and services.
3. Service orientation in architectures.

Further, business process developments related to system implementation and the respective cost effects were discussed. Views about software market evolution and the technologies, integration and technology evolution were also brought up.

These areas were considered, to detect possible explanations for the findings in the software "make or buy" decisions and, in the latter case, the selection of the supplier when acquiring systems and applications.

There was a requirement that persons to be interviewed should occupy positions involved in the respective decision making on software development, application portfolio management and software (license) procurement. These positions vary from CEO to IT-architect. Each interview was recorded and transliterated for further analysis. The analysis of the qualitative data follows the so called template approach derived from (Robson 2000, p. 458). The interview topic scheme is derived from the general software market evolution model, and the software selection criteria are used as a supporting theory. The interviews in Europe were conducted by Dr Lauri Frank, Prof. Pasi Tyrväinen, and in some cases together with other members of the SmarTop project group. The interviews in Asia were conducted by Prof. Jari Veijalainen and Caleb Zhao who acted as an interpreter in all Asian interviews. These interviews were conducted in 2007–2008. The interview scheme remained the same in all interviews.

1.2 Interviewed Operators

The CSP companies that were interviewed cover an area depicted in dark gray in the map (see Fig. 1). The geographical area of operations of these telecom

Fig. 1 The area of operations of the interviewed operators shown in dark gray

companies covers Northern Europe, Eurasia, and parts of Middle and South Europe (North American market is rather well covered with the quantitative data analyzed in the subsequent sections of this chapter, and hence no interviews with the operators in that market were conducted). The operators interviewed Europe are classified as "Regional" if they operate only within a regional area in one country, as "National" if they have activities in one or two countries, and as "International" if they operate in several countries.

Since the rapid economic development in Asia is maybe one of the greatest factors in the current CSP and related software business, a second wave of interviews was conducted with Asian operators in mainland China and in Taiwan. The operators interviewed in Asia fall into two categories: National or Regional, in general larger in size than those in Europe. Due to demographics, Asian operators typically serve a larger customer base than do European operators.

In the following tables are the details of the companies where the interviews were conducted. Table 1 summarizes the information on the European operators, and Table 2 the operators interviewed in Asia. The information on the business was gathered from company websites and annual reports.

1.3 Changing Telecom Business Environment in China

In June 2008 the Chinese government decided to reorganize the country's telecom market. The situation during summer 2008 before the reshuffle is described in Table 3 below. It is worth noticing that all old and new operators are mainly owned by the Chinese state, which owns, for example, ca. 80% of China Unicom (China Unicom 2008).

Table 1 Interviewed companies in Europe

CSP type	History of operator	Current company establishment	Main access networks provided	Further access services	Subscribers
Regional 1	Regional operator since the late 1800s	Concern structure with several daughters	PSTN, ATM, cable WiMAX (wireless broandband)	Shareholder in a mobile operator company, owns servicing companies	<0.1 mio
Regional 2	Regional operators since the late 1800s or early 1900s	Major mergers in the mid 1990s and in 2000s	PSTN, cable	Shareholder in a mobile operator company	<0.1 mio
Regional 3	Regional operator since the late 1800s	Major mergers in the mid 1990s and in 2000s	PSTN, cable	Shareholder in a mobile operator company	<0.1 mio
National 1	Regional operator since the late 1800s	Series of mergers, established early in the 2000s	PSTN, cable, mobile	All services	<10 mio, >1 mio
National 2	Incumbent national operator since the mid 1800s	Established in the late 1990s	PSTN, cable, mobile	All services	<10 mio, >1 mio
International 1	Incumbent national operator since the late 1800s	Established in the late 1990s	Mobile	In a holding company together with PSTN/ cable provider	>10 mio, <100 mio
International 2	Incumbent national operators since the late 1800s	Merger early in the 2000s	PSTN, cable, mobile	All services	>100 mio

Table 2 Interviewed companies in Asia

CSP type	History of operator	Current company establishment	Main Access Networks Provided	Further access services	Subscribers
National 1 China	A national operator since 2000	?	GSM network and 2.5 G services		ca. 400 mio
Regional China	Regional operator since 1994	Belongs to national concern structure with several regional daughters	GSM network and 2.5 G services		< 20 mio (the whole concern ca. 130 mio)
National 1 Taiwan	Mobile operator since 1998	Series of mergers during the 2000's	PSTN (enterprise) Cable: TV, Broadband Mobile	All services	< 0.3 mio < 0.6 mio ca. 6 Mio
National 2 Taiwan	Established as a mobile operator	Established in 1996, several mergers since	Mobile Broadband (SPARQ)	Mobile, Broadband	ca. 6.3 mio ca. 0, 3 mio
National 3 Taiwan	The incumbent operator, established as corporation July 1, 1996	Expanded to mobile and broadband	Mobile (GSM, UMTS) PSTN Broadband (ISP) IPTV	All services	9 mio (>3 mio UMTS subs) over 10 mio 3–4 mio 1 mio

Table 3 Chinese telecom market in May 2008 (Jeung 2008a)

Operator	Total assets (billion Yuan)	Total users (M)
China mobile	700	392
China telecom	254.9	255
China netcom	186.4	131
China unicom	72.6	162
China tietong	55.4	21

The table shows that the total number of subscribers approaches 1 billion. That number covers the subscribers of both wireless and wired networks. It is not clear how much overlap there is among the segments.

Table 4 contains an overview of the planned market structure with three new full-fledged operators. In the new structure all new operators have mobile, fixed-line and broadband business. The 3G technologies will be allocated to different operators in the new setting, whereas two of the new operators will operate a GSM network. The TD-SCDMA network, based on home-grown 3G technology, will be operated by the new China Mobile. It will be formed from the current China Mobile

Table 4 Chinese telecom market after the revamp (Jeung 2008a–d)

	China mobile (China mobile+China tietong)	China telecom (China telecom+ China unicom CDMA part)	China unicom (China unicom GSM+China netcom)
Mobile subscribers (M)	399 (GSM; TD-SCDMA)	43.09 (CDMA; CDMA 2000)	129 (GSM; WCDMA)
Fixed-line subscribiers (M)	20.79 (China tietong)	216	107
Broadband subscribers (M)	4 (China tietong)	38.36	24.5

and China Tietong, a fixed-line operator. CDMA and CDMA 2000 networks will be operated by the new China Telecom. The company will be formed from the existing wired operator China Telecom and the CDMA network and its supporting organizational structures of China Unicom. Finally, the new China Unicom will be formed of the old China Unicom and China Netcom. The former is a combined GSM and CDMA operator and the latter is a wired operator with a substantial fixed-line and broadband customer base (over 100 million). This operator will operate a WCDMA network. The 3G licences are to be distributed to these new operators as soon as the restructuring has been completed. The latter phase was reached in November 2008 and the 3G licenses were issued in January 2009, as anticipated in 2008 (Jeung 2008b, Jeung 2009a). The restructuring has been achieved by selling entire companies or their parts to other actors. Thus, the CDMA network and related organizational infrastructure of China Unicom, was sold to China Telecom in October 2008 for 43.8 billion yuan. China Tietong was sold to China Mobile in October 2008.

There were 42 million CDMA subscribers in China at the end of October 2008. The target is to have 100 million CDMA, CDMA 2000, and CDMA 2000 EV-DO users by 2010. From the infrastructure viewpoint, China Telecom has 100,000 CDMA base stations and China Unicom had about 320,000 base stations in its network that was sold to China Telecom in October 2008 (Jeung 2008d,f). China Unicom has 120,000 GSM base stations in its current network (Jeung 2008f). The market situation after the reorganization is unbalanced in that China Mobile has the largest overall market share. The Chinese regulator might therefore introduce measures to balance this situation. Suggestions have been made, e.g., to lower the network interconnection charges and introduce number portability (Jeung 2008e). In addition, the Chinese government ordered in October 2008 that the new operators have to open their base station towers and trunk network to their competitors (Jeung 2008f). This decreases the capital expenditure in building the 3G networks. It has been estimated that the overall capital expenditure used to enhance and upgrade the networks in the mainland China will be over 100 billion yuan (over 10 billion euros) in the coming few years (Jeung 2008f).

All in all, the interviews were conducted at the time the new market structure was already in sight, but the participants did not yet have a clear understanding what changes would be caused by this in terms of the software production and usage that interests us in this context. One can argue, though, that the new operators need more knowledge from outside than in the previous phases, because the schedule to start 3G network operations is very tight. Software-wise they are most probably forced to rely heavily on external software providers.

2 CSP Software Acquisition Strategies

2.1 Software Development vs Application Procurement

The developments in the application portfolios of the providers is tied to their technology portfolio management on one hand, and on the other, to their business portfolio, i.e., the variety of services the CSP chooses to provide. New services or a change in the existing ones normally mean some change or new developments (maintenance measures or software development and/or integration) in the set of OSS/BSS systems and applications the company is maintaining. Besides technology evolution, business development drives the developments in the OSS/BSS software and systems. Resources are required besides developing or buying software licenses, also for the maintenance of existing systems. With maintenance, we understand, as defined with the ISO/IEC 14764 standard, the corrective, adaptive, perfective and preventive measures, modifications and improvements of systems and software in the use of an enterprise.

For its existing and new OSS/BSS software systems, the CSP may consider whether to develop and maintain the new application or system using its own resources, or whether to buy a service to tailor the system specifically for them, or to acquire a license for a commercial off-the-shelf software component (COTS). The general model of software market evolution predicts a development where the internal software development in the industry enterprises disappears, and the portion of COTS licenses as well as the spending on software related services is increasing.

The interviews were thus looking at the proportions of software development in-house, and the procurement of OSS/BSS software and related services (like integration projects) by considering two issues:

1. To what extent and for what reasons do CSPs develop software in-house; and
2. If software is acquired from external software vendors, what are the reasons a CSP uses for selecting its OSS/BSS software providers. With this, also the aspects of bespoken software or COTS are discussed with the interviewees.

This exploration does not look at the application services provisioning (ASP) or software as a service (SaaS) schemes particularly. Under the rubric *services* go mostly (according to the interviewees) integration of systems and maintenance related services. ASP and SaaS, if available, are included.

The first issue above is related to the general model of vertical software market evolution (Tyrväinen et al. 2008; see Chapter 4 "Model for Evolution of a Vertical Software Industry" in Part I above), and the second issue is reflected on a common vendor selection criteria classification (Kotler and Keller 2005), which is briefly presented further below. The selection of suppliers is used as complementary information on the software procurement in gauging the status of the market segment. The empirical analysis of these issues follows, reflecting the material from six interviews with European CSP representatives, and from further five interviews with representatives of Asian CSPs, on these theories.

2.2 Criteria for Supplier Selection

With the maturing of an industry specific software market the industry companies begin acquiring software from outside; also the supplier considerations become now important. To complement the analysis of issues affecting the market developments, we will also look at the supplier selection criteria the interviewed CSP representatives mention. In previous research, different categories of purchases have been found according to the supplies being acquired (Kotler and Keller 2005): (1) routine-order supplies (supplies like raw material for a production process, office supplies etc.), (2) procedural-problem products (acquiring of such product can cause a change in a production process, or influences the efficiency etc.) and (3) political-problem products (these products have a deep influence to the business, requiring a process or procedure change). Acquiring software for an operator enterprise seems to fall into category (2) for smaller scope applications or (3) for large systems.

The most important criteria for the supplier selection in category (2) purchases according to the classification (Kotler and Keller 2005), are

- Technical service
- Supplier flexibility
- Product reliability.

For category (3) purchases, the most important supplier selection criteria are

- Price
- Supplier reputation
- Product reliability
- Service reliability
- Supplier reliability.

Finding other criteria in use for supplier selection when acquiring software for the CSP, will point to special requirements that this industry companies set to the software support for their processes, or other factors. If such additional criteria or non-conformance to the normal criteria are found, this may – together with a possible deviation from the general model of software market evolution – help to understand the developments around the CSP industry.

2.3 *Software Acquisition – Make or Buy*

On software acquisition (topical areas 1 and 2), the operator representatives were asked how they acquire the software they are currently using in their operations and business, and why they think they should or should not be involved in the software development business. A further question was about how they see the future of software development in this industry domain. The market is occupied by different types of vendor companies. There are industry-specific vendors developing telecom specific software, and large vendors providing also hardware equipment, or large renowned software vendors. There are also smaller young companies providing solutions for a limited scope. The interviewees were asked how they see the role of the different providers, and their own role in the developing of the applications for their business operations and new software systems in general for the communication services provisioning business. Questions about criteria used in the selection of

Table 5 The CSP software development/procurement: European operators

CSP	Software developed in-house	Software purchased: licences, projects from vendors	Current status of SW development	Issues and trend in software development/ acquisition
Regional 1	<10%	>90%	Some work on integration	No development, no interest in maintenance
Regional 2	10%	90%	No significant development	No development
Regional 3	10%	90%	No significant development	No development
National 1	15%	85%	Maintain legacy, mostly no development, an in-sourced developing unit	Design innovative software, more partnerships
National 2	40%	60%	Has a spin-off, design software and architectures	No own development, keep design and software IPR
International 1	<20%	>80%	No own development, design software important	Evolve to a design-house
International 2	50%	50%	Has outsourced development own development for (1) integration of own systems (2) cutting of costs (3) innovation	Not giving up all design and development, innovation

Table 6 The CSP software development/procurement: Asian operators

CSP	Software developed in-house	Software purchased: Licences, Projects from vendors	Current status of SW development	Issues and trend in software development / acquisition
National 1 China	< 5 %	95%	No significant development	
Regional China	< 5%	> 95%	Only tailoring[1]	Increasing outside vendor usage[2]
National 1 Taiwan	50%	50%	Development department. with 500 people[3]	Knowledge about architecture and processes they would not like to outsource
National 2 Taiwan	20 %	80%	Development dep. with 200 people	Use also big foreign software companies[4]
National 3 Taiwan	40 %	60 %	A lot of in-house dev. 1300 people in the own lab	100–200 small apps. from service prov.

[1]This operator has a few software engineers from a local software company at their premises performing tailoring and other tasks for them continuously on request.
[2]This operator uses 70–80% of their investment budget for hardware, about 10% for software purchases and 10% for personnel.
[3]This operator uses of the their software systems expenditure 25% for the equipment, 40% for own personnel, 30% for software.
[4]Overall software systems expenditure 40% personnel, 40% HW, 20% SW (10–15% of HW price is SW)

the software vendor were also asked. These questions were open-ended so that other possible issues could also be brought up.

In Tables 5 and 6, the basic data on the topic areas 1 and 2 is presented: the type of telecom operator, their estimated percentage of spending on internal software development based on the current accounting data, the estimated percentage of spending on software licenses (software products) or services (i.e., for software projects conducted by an external provider). The comments on the current status and the development trends in the software development as compared to acquisition of licenses for the foreseeable future are listed in the last two columns.

The situation on the Chinese telecom market is still, mid of 2009, not fully settled. The Chinese government decided in June 2008 to revamp the telecom market by 2009. Of the previous five operators, only three emerged after merging operations in late 2008. Further, the government has assigned the 3G technologies to each of the three new operators. As of the time of the interviews, (July 2008), the 3G licenses had not been issued, but this was scheduled to happen in few months. The changes on the Chinese telecom market are radical to such an extent that the past practices have to be changed, if not abandoned. Our view is that the operators are forced to buy more knowledge and software from outside, especially from those companies that have experience in running simultaneously 2G and 3G mobile networks and fixed and mobile networks.

All in all, in the overall development, the telecommunications industry seems to be following the general pattern of software market evolution within a specific industry. This is confirmed by past developments indicated by the interviewed CSPs: they have legacy software systems developed in-house still in production. Also, some of them have spinned off their software development to an independent company that thereafter can develop and offer OSS/BSS software for a broader range of customers. Another way is to move that development to a daughter company.

With the European operators, the pattern of the general model of software market evolution seems to be accompanied with slight differences between different company size categories. Smaller operators, due to their limited resources, have outsourced almost all software development and rely on external vendors also in further process automation and enhancement. The middle sized CSPs hold on to some software development, and are willing to own the software designs of new applications developed to create some novel types of services. However, they seem to have no plans to include some software offering in their business portfolio. Software design in-house is for them a means for remaining competitive with some possible new services offering, which they attempt to design themselves, and retain the IPRs for the related software.

In Asia, the situation is different in mainland China compared to Taiwan, Taiwanese operators being of a similar size as the middle category European ones. With the Chinese, there is a clear difference in the scale of operations, but this does not seem to influence the evolution of the software development activities. The CSPs in mainland China seem to follow the pattern without major deviations. They gradually abandon in-house software development and look for external vendors to provide the solutions they need. The CSPs in Taiwan seem to divert to an extent from the suggested evolution pattern in a similar way the European middle-sized companies do. The availability or need of some resources seems to lead to this:

- Maintenance of legacy software requires some work. There are still bespoken systems for which the maintenance expertise might not be easy to find.
- Since staff with development skills is available for maintenance tasks, that human resource is used also for some application development.
- Further, it is important to retain knowledge on architectures and processes in-house to ensure competitiveness in the future.

The special feature of the Taiwanese telecom market is the important role of a call center. All operators invest a large portion of their resources into it (e.g., one of them has 100 people in their call center). This is also an activity they would not like to outsource. There was some interest in retaining software development as part of the business: one of the Taiwanese operators has initiated discussions about selling some systems to other South-East Asia countries, such as Vietnam and Myanmar.

In the quantitative studies conducted, it was found that the large European and North American operators also still use a significant amount of resources to software development – the reasons for this can not be read, however, directly in the data.

For those operators that have outsourced software development, there clearly is a threshold to go back into developing software in-house. These very large Chinese or small European ones in the group of studied operators are likely to be a market for different providers. The middle category will be a partner also in software industry.

The software for support processes is of little interest, however, for any of the operators. Both the small and the larger operators seem to welcome licensed software packages for enterprise resource management, accounting etc. Telecommunications-specific ERP systems did not seem to be a hot topic in the interviews. In Asia there is some more emphasis on the preference for comprehensive solutions. This may mean that the companies there are entering the business at a more mature stage. In Europe, the enterprises are going through a longer development path from silo systems through system integration to service-oriented architecture (SOA) solutions and enterprise systems.

The middle category of operators seems globally aiming to be a trendsetter in the development of value added services. In Europe, there were decisions against the regular vendor selection criteria. Operators are willing to take risks in vendor selection to find innovative solutions for creating new business. The large operators with broad geographical range of operations and very high numbers of customers have more weight in the decisions in which technologies will become prevalent. As the example of China indicates, there can be nation-wide decisions on technologies that do not remain without impact internationally.

A strong driver for software designs is currently services orientation, with which the business and the system developers are finding common ground. This was therefore taken up as a topical area in the interviews. The deployment of unifying platforms for services development is under consideration in all interviewed organizations, if not in progress already. A service delivery platform or framework (SDP/SDF) seems, however, to be a concept still emerging. There are commercial reference architectures and marketing concepts of SDPs, but a unified reference model does not exist. The own efforts of the companies focus mainly on service delivery frameworks (cf. application frameworks). Further, the variation of services and the approaches to the development of customer services are areas of strategic consideration within the CSPs.

In Asia, no small operators were interviewed. The company size and numbers of subscribers there are generally larger than in Europe. It looks like the very large operators of China are also outsourcing all software development. Interest in software designs that resembles interest among the middle range European operators can be found in Asia with the middle range operators. Furthermore, good quality customer care is with the middle category of Asian operators a competitive factor – this may be partly due to the difference in the cost of work force in Asia as compared to the high cost levels of Europe. The history of the companies in Taiwan is, due to the political environment in that part of the world, different from China, and the Taiwanese businesses show similarities to the Western private businesses. It looks like the commercial interest and the entrepreneurial strategy building with good customer care and development of attractive value added services is as prevalent in Taiwan as it is with the middle range European CSPs.

Analyzing the Current Phase of the OSS/BSS Software Market

Lauri Frank, Pasi Tyrväinen, and Eetu Luoma

In this section, the evolution of the OSS/BSS software industry is described, and the current phase of the OSS/BSS software market is analyzed using quantitative data. The first analysis uses the viewpoint of the general industry lifecycle stage models. This is followed by an analysis from the viewpoint of the vertical software market evolution model. The last section analyses the mergers and acquisitions of the OSS/BSS market from the last few years.

1 Industry Structure Evolution

A majority of CSP's who have traditionally themselves developed and maintained all the software needed to run their operations, recently have adjusted their business strategy to concentrate on their key competence area, communications. Thereby they have also outsourced software development and maintenance to the newly emerging OSS/BSS software industry. Sanz et al. (2006) from IBM note that telecommunications operators have modularized their operations in the form of an agreement on the enhanced telecom operations map. This modularization and standardization of operations is the result of telecom operators' willingness to exploit the economies of scale arising from external providers supplying software to the telecom operators' common needs.

Our interviews with six telecommunications operators and four OSS/BSS vendors in Europe indicate that, until approximately the 1980s, voice services over fixed telephone lines were the only ones provided by the CSPs to their customers. The operators were mainly national incumbent firms or state agencies to begin with. All the operations for providing these services were conducted within

L. Frank (✉)
Department of Computer Science & IS, University of Jyväskylä, P.O. Box 35, 40014, Finland
e-mail: Lauri.Frank@jyu.fi

P. Tyrväinen and O. Mazhelis (eds.), *Vertical Software Industry Evolution*, Contributions to Management Science,
DOI 10.1007/978-3-7908-2352-3_9,

the CSP and thus were vertically integrated. The needed operations software (OSS) was developed within the CSP, unless it came with hardware equipment such as with some switchboard solutions by companies like Ericsson and Siemens.

In the 1990s, the situation changed: on one hand, the regulation was rescinded and the markets were liberalized, and on the other, new communications technologies emerged and were added to the CSP's service offering list. Besides voice communications, CSPs now started to provide, for example, broadband internet connections and mobile communications services. It should be noted that each of the services was usually operating on the top of its own specific network technology and its own OSS/BSS.

Later on, some of the OSS/BSSs have been unified across the CSP's services. For example, CRM systems could be the same for all the provided services of the CSP. Also, some of the provided services have been integrated, for example, broadband or voice can nowadays be provided over several types of networks. Moreover, some telecommunications operators have specialized, by focusing only on a part of the traditional telecom operators' functions; this has resulted in the appearance of, e.g., service or network providers.

The interviews with the telecom operators and discussions with software vendors suggest further that in the near future, operations may be further integrated over the supported technologies, and that the number of network technologies to be supported would decrease, partly due to phasing out of PSTN networks and abandoning copper cable altogether. It seems to be already the practice that in new areas opened for construction building, only fiber cable will be part of the community infrastructure. In less densely populated areas, the use of wireless broadband technologies (e.g., HSPA or WiMAX) is the alternative.

Also, nowadays, network access and operations, apparently, could be acquired as managed services, leaving the provisioning of connection to the customer as the only activity for the CSP. This evolution path of the CSPs, from vertically integrated to horizontal, is illustrated in Fig. 1.

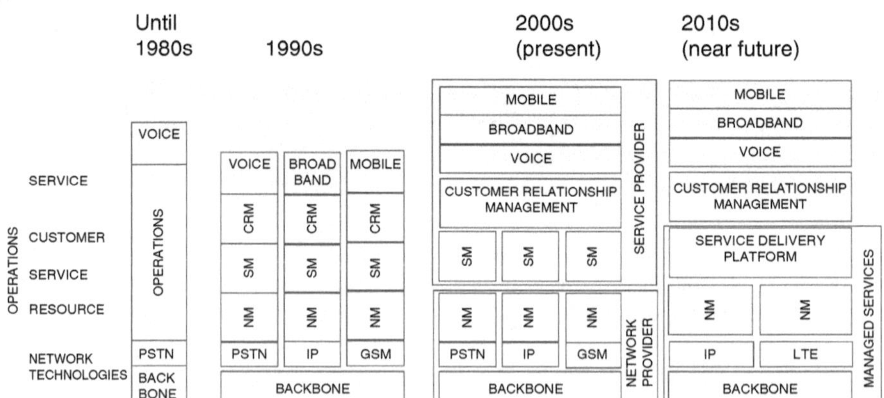

Fig. 1 The evolution of the industry structure for a telecommunications operator (CSP)

As Fig. 1 shows, operations have previously been an integral part of the operator's business. The software supporting the operations was also produced by the operator in-house. Gradually, however, these software development and maintenance tasks have been outsourced to external software firms and system integrators. This can be considered as a vertical disintegration of the telecommunications operating industry.

CSPs offer already voice and broadband services horizontally across multiple physical network layers. Also the customer relationship management applications have adopted a horizontal approach, where the same software is used across multiple network technologies instead of providing separate customer interface systems for PSTN, mobile and broadband, as was the case in the 1990s. The near future horizontalization trends are related to the network and service layers. In the network layer, operators seem to want to get rid of multiple network technologies and converge into two dominating technologies, namely all-IP and LTE. All-IP approach aims at adopting the IETF protocol stack used in the Internet also as a general communication service platform. Using ip-phone applications for voice communication (such as Skype) would bring down CSP revenue and enforce adoption of flat monthly fee charging. The CSPs are somewhat unwilling to adopt this path as long as no replacing revenue streams and business models have been identified. Applying all-IP approach to the mobile world (e.g., in form of WIMAX) has also some technical problems related to standardization and roaming. Thus the mobile operators have mainly adopted an alternative approach, long term evolution (LTE) which secures CSP investment on mobile infrastructure and provides a migration path to mobile broadband. In addition, use of SIM-card based customer lock in is still possible in the LTE architecture but not in WIMAX, mobile WLAN and other pure IP approaches.

In the service layer each of the network technologies has traditionally had separate tools for service provisioning, assurance and billing. This has caused multiplication of work and tools in service creation and other OSS/BSS processes. There have been approaches to use single network technology independent tools for service creation and execution, such as IP multimedia subsystem (IMS) and recently service delivery platforms (SDP) and service delivery frameworks (SDF). The goal of service delivery platforms is to create a unified environment for creating new services and for executing them regardless of the network technology used. The service delivery frameworks are not as ambitious and do not offer a uniform environment; rather they offer a set of standards, principles, and policies guiding the creation and management of services throughout their lifecycle (in essence, the SDF is an embedded part of a unified SDP). This bold idea has not been realized to a large extent yet, and thus the term service creation framework is used more widely to describe a set of primitive supporting services (billing, activation etc.) which can be activated as a part of a new customer service from a service creation / execution tools using a backplane or middleware software. In any case, there would be a market for SDP/SDF environments if implementing such would be feasible. If such environment would appear, it could form a dominant design to which all other OSS/BSS software components would have to integrate.

During the SmarTop project, the interviewed mid-sized communication service providers indicated that their companies are concentrating their software acquisitions on only few, larger providers, which can offer lower administrative and integration costs. Meanwhile the software vendors aim to expand their range of supply to meet the demand, through introducing new product lines and corporate acquisitions. These can be seen as signals reflecting the consolidation of the telecom software market, and the notion motivates examining the current phase of the market as well as the likely subsequent developments. In the following, the structure and the structural changes of the telecom software industry are analyzed. The aim is to analyze the evolution of the telecom software industry following the theories and models of industry evolution.

2 OSS/BSS in General Industry Lifecycle Stage Models

The models in contemporary research literature can be used to model software industry evolution. The models include the industry lifecycle theory (Gort and Klepper 1982) and the theory of vertical specialization and integration (Macher and Mowery 2004). An analysis using the industry-specific vertical software market evolution model (cf. chapter 4 on "Model for Evolution of a Vertical Software Industry") will be presented in Sect. 3. These theories supported by some empirical observations, can clarify the following stages of the development of the telecommunications industry and the vertical software market.

A. *Early stage*. Traditionally, CSPs have internally developed and maintained the software required to run their operations. Until the 1980s they only provided voice services over fixed telephone lines and at that time the telecommunications operators were largely national incumbent firms. All the necessary operations were conducted within the CSP, that is, the operations were vertically integrated. The needed operations software was developed within the CSP for its specific purposes and network infrastructure.

In the 1990s, the markets were liberalized and new communications services, including mobile communications were added to the CSPs service portfolio. Still, the services were mainly operating on top of their own specific technologies and operations supporting systems. By that time, CSPs started focusing more on their key competence area, i.e., communications, and consequently outsourced software development and maintenance to software vendors or ICT service providers. As a result of this, a new software market was born.

In the first stages of the OSS/BSS market, only a few firms were operating in the market, which is in line with the mentioned theories and models. In the case of telecom software, these firms were typically founded by outsourcing the operators' software production. These few firms provided software for most of the operators' operations. The OSS software industry was therefore vertically integrated and relatively consolidated as an industry.

B. Growth of the industry. After the initial stage, new vendors providing more specialized software solutions emerged. These vendors disintegrated the telecommunications software market vertically and fragmented the market structure. As a result of the increase in the number of vendors, the OSS software industry became more vertically disintegrated and competitive. From the demand side, the outsourcing of non-core business operations also seems to have driven the vertical disintegration. These developments are also in line with the above mentioned industry studies with regards to a growing number of vendors and a less consolidated market structure. The key reason for this vertical disintegration seems to be the desire to gain efficiency, innovativeness and flexibility in software manufacturing as mentioned above by Macher and Mowery (2004).

C. Industry matures. Later on, most of the CSPs' operations have been unified on a conceptual level, with telecommunications operators having modularized their operations in the form of agreeing with the Enhanced Telecom Operations Map (Sanz et al. 2006). Thus, software manufacturing can be considered as been vertically disintegrated from the operator's other functions. Some of the services provided by a CSP have been integrated and some telecommunications operators have specialized as service or network providers. In the near future, operations are likely to be further integrated over the supported technologies, and the number of technologies is likely to decrease partly because of, for example, the disappearance of the PSTN network. Also, it seems that network access and operations could largely be acquired as managed services, leaving the connection to the customer as the only area of activity for the CSP.

In the latest stage, the OSS software industry shows indications of reconsolidation and also of vertical reintegration. On one hand, this seems to be because of the vendors' active business strategies of conquering a larger share of the market (see also Frank et al. 2007). On the other hand, this development arises from the customers' (the CSPs') aim to concentrate their procurement to only a few software vendors each. Thus, in this stage, the number of firms should decrease.

The theories and indicators related to the described stages are summarized in Table 1. According to the industry lifecycle theory, the number of firms competing within an industry increases in the growth stage and declines in the maturity stage. Conversely, the concentration of the industry, measured with the HHI index, decreases in the growth phase and later, in the maturity stage, raises again. With regards to measuring the extent of vertical integration, we suggest examination of different software vendor strategies, namely their breath of operations (cf. Frank et al. 2007). In the growth phase, the industry has more specialized niche companies,

Table 1 Stages and indicators of the OSS/BSS software industry evolution

Stage	Early	Growth	Maturity
Theory (indicator)			
Industry lifecycle(number of firms)	Low	High	Low
Vertical integration(breadth of operations)	High	Low	High
Consolidation(HHI index)	High	Low	High

whereas in the maturity stage there should be more companies providing software for the entire vertical process area or broadly for multiple segments.

The market data provided by Dittberner Inc. enables us to investigate the evolution of the mentioned indicators from the years 2000 to 2006. The OSS industry was growing rapidly during the mid-1990, and we assume the industry to have passed the first stage. Thus we hypothesize the industry having started from the second stage (growth) and shifted to the third stage (maturity) within the observed time frame. For this shift between stages two and three, the following hypotheses can be derived from Table 1:

1. The number of firms decreases as the industry reaches the mature phase of the industry lifecycle.
2. The breadth of operations of these firms increases for the above mentioned reasons.
3. As a result of the two above hypotheses, the industry consolidates, as fewer vendors have a greater market share of the industry.

In order to examine the concentration of the software market, commercial data from Dittberner Associates, Inc was analyzed for the years 2000, 2002, 2005 and 2006. The OSS/BSS companies and their revenue data were organized into thirteen segments presented in Fig. 1 of Chap. 4. The number of companies in each segment, the HHI index value for the industry segments, and the number of companies deploying vertically integrated business strategies for each of the year data was provided. The HHI is calculated by summing up the squares of individual firm market shares from all firms operating in the market. It thus gives a greater weight to firms with greater market shares. The mathematical form of HHI is:

$$HHI = \sum_{i=1}^{n} S_i^2 \tag{1}$$

In (1) S_i is the market share of Firm i, and n is the number of firms in the market. The value of HHI is thus positive, and its maximum is 10,000 when a single firm has a market share of 100%, which is the case in a monopoly situation.

The total number of companies operating in the market at different points in time is shown in Table 2. Table 3 represents the number of companies operating in the thirteen segments and the changes in the number of vendors for the time periods. A single vendor may have operated, i.e., received revenue, from one or several segments. In general, the number of vendors operating in the industry (including service providers) dropped from its peak of 232 companies in 2002 to 196 companies

Table 2 The total number of analyzed companies operating in the market (years 2000–2006)

Year	# Companies
Y 2000	223
Y 2002	232
Y 2005	207
Y 2006	196

Table 3 Changes in the number of companies in the OSS/BSS market segments in the years 2000–2006

Segment	1	2	3	4	5	6	7	8	9	10	11	12	13
	N	N	N	N	N	N	N	N	N	N	N	N	N
Y 2000	36	34	27	51	24	36	35	25	29	36	17	49	25
Y 2002	32	33	28	48	28	35	40	26	32	37	17	49	27
Y 2005	26	21	27	34	29	31	30	22	26	31	17	41	31
Y 2006	28	18	27	33	25	28	30	22	25	29	16	42	33
dN%	−22.2	−47.1	0.0	−35.3	4.2	−22.2	−14.3	−12.0	−13.8	−19.4	−5.9	−14.3	32.0

Table 4 Changes in the HHI index in the OSS/BSS market segments in the years 2000–2006

Segment	1	2	3	4	5	6	7	8	9	10	11	12	13
	HHI	HHI	HHI	HHI	HHI	HHI	HHI	HHI	HHI	HHI	HHI	HHI	HHI
Y 2000	1,371.8	908.8	866.1	1,114.6	2,985.3	747.2	1,391.1	2,105.2	612.0	704.1	1,759.4	729.6	1,917.0
Y 2002	1,480.9	1,041.5	796.9	1,676.8	1,547.4	941.7	977.8	1,929.4	595.5	595.1	1,876.9	699.3	1,535.7
Y 2005	2,126.5	1,701.6	1,021.8	2,363.2	777.4	1,235.1	1,078.4	1,689.0	630.4	747.0	2,040.6	771.4	1,368.8
Y 2006	2,209.0	1,441.4	861.6	2,777.7	1,042.5	1,827.3	801.0	1,657.3	716.3	1,006.2	1,880.7	703.6	1,259.6
dHHI%	61.0	58.6	−0.5	149.2	−65.1	144.6	−42.4	−21.3	17.0	42.9	6.9	−3.6	−34.3

in 2006. Also, the table clearly shows that in almost every OSS/BSS product segment, the number of vendors has decreased. We speculate that one main reason for this decline is the mergers and acquisitions carried out during the observed time period.

Table 4 depicts the development in the market concentration of the OSS/BSS segments as measured by the changes in the HHI index values. The findings are mostly in line with the analysis of the number of companies and our hypothesis regarding the phase in the industry lifecycle model. In seven segments, the HHI index value has increased, up to 149.2% in the Offline Billing segment and in six segments it has decreased. The segments of "Fraud Management and Revenue Assurance" and of "Online Charging" show major negative values in the change of the HHI index. This could be explained by these comparatively new product segments being in a different evolutionary stage within the OSS/BSS industry. To summarize, while the number of companies is falling, the industry becomes more concentrated. This indicates a shift from the growth stage to the maturity stage in terms of Table 1.

In order to analyze the vertical integration of the OSS/BSS companies, they were first classified into categories applying the following rules (see Fig. 1 of Chap. 4 for the market segments, vertical and horizontal layers). A company was treated as carrying out a niche strategy in case the vendor received revenue from only one of the thirteen segments. A company executed a vertical strategy if it had revenues from two or more segments in different layers, but only on one of the verticals (fulfillment, assurance or billing). A company had a broad strategy in case it was present in multiple verticals and horizontal layers. The service providers were also included in this classification as a separate category. Table 5 presents the percentage of the companies on each strategy differing by the breadth of operations and the changes in the percentage of companies with differing breadth of operations within the observed time period. The results show a minor shift towards an increasing breadth of operations because of a higher decrease of a niche strategy companies compared to broader strategies. As also service providers may be considered as providing customer specific solutions for various segments, and therefore could be treated as executing a broad operations strategy, one could draw a conclusion supporting the suggested hypothesis.

We suspect the telecom operators will concentrate their procurement even more to only a few providers and telecom software vendors to broaden their operations. This shift will cause a concentration in the OSS/BSS market, which is in line with the latter stages of the previously mentioned industry evolution theories. This so called 'industry shakeout' has been present in the observed industry: the number of OSS/BSS software vendors has decreased between the years 2000 and 2006.

Table 5 Changes in the breadth of operations of the OSS/BSS companies in the years 2000–2006

Breadth of op.	Niche (%)	Vertical-broad (%)	Service providers (%)
Y 2000	30.5	43.5	26.0
Y 2002	30.4	43.2	26.4
Y 2005	29.0	38.6	32.4
Y 2006	27.0	39.8	33.2
d%	−11.3	−8.5	27.5

Also, the industry concentration models' prediction of increasing concentration was found to exist in the OSS/BSS software market.

In summary, the empirical data suggests that, according to the industry lifecycle model, the extent of vertical integration, and the market consolidation, the OSS/BSS software market is shifting from the Growth phase to the Maturity phase. To conclude, the empirical data would indicate that the OSS/BSS software markets are reaching a mature phase, where one could expect a more stable development afterwards.

On the other hand, a new cycle of vertical disintegration and de-concentration could occur in the future. The results also indicate that the market segments of the OSS/BSS software industry might be in different evolutionary stages. If the industry is considered as a whole, it may be judged to be in the mature phase. However, six market segments indicate an opposite development from the rest of the industry by, for example, showing a stable or even growing number of firms. Within an industry there thus seems to be segments which evolve differently. This might be due to the different ages of these segments; relatively newer market segments could be expected to behave differently from the phase the rest of the industry is in.

3 OSS/BSS in the Vertical Software Market Evolution Model

A vertical software market can be analyzed with the vertical software market evolution model and the industry lifecycle model. Usually the market is identified to be in similar phases according to both of these models. That is, the market is found to be in growth phases in both, in mature phase in both, or in between in both models. However, in some cases the vertical software market evolution may be stopped causing the software vendors to allocate their resources to some other market with higher expected revenues and thus causing transition to a mature phase in terms of the industry lifecycle model. Thus the previous result suggesting that most of the OSS/BSS market segments are in the mature phase of the industry lifecycle model does not necessary imply that the vertical software market would as well be in the last (mature) phase. The current phase of the vertical software market evolution manifests itself in a number of different indicators, among which one can distinguish software spending by the operators, the channel of software delivery, market concentration, etc. Some of these indicators are analyzed next.

3.1 Software Spending by CSPs

As mentioned above, the current phase of the software market evolution, among other indicators, is reflected in the software spending by the CSPs. According to the model of the vertical software market evolution described in Chapter 4 on "Model for Evolution of a Vertical Software Industry," software spending is high in the first

Table 6 Software spending by CSPs, in relation to revenues

Phase of vert. software market evol.	Innovation	Productization	Adoption	Service
Total IT spending (%)	<5	5–15	~2	<2

Table 7 The software spending (minimum, maximum, average, and medium values) of CSPs in relation to their revenue

Year	Min	Max	Average	Median
2005	0.00	0.14	0.038	0.028
2006	0.00	0.20	0.039	0.026
2007	0.01	0.10	0.031	0.029

and especially in the second phase of the evolution (where it may reach and exceed 10% of the revenue), while in the subsequent phases the spending drops to 1.5–2% (Table 6).

In case of CSPs, software spending consist of in-house software development costs and the costs of acquired software products and services. Based on the CSP's annual reports and assuming that the in-house software development is a portion of the CSP's R&D spending, Table 7 shows the software spending of a sample set of CSPs, in relation to their revenues, for the years 2005–2007. As can be seen, the total software spending in the period 2005–2007 is around 4% (0.031–0.039) and in some cases exceeds 10%. This rather high value indicates that the software market has not reached the third phase yet, and hence is currently in the second phase (Productization and Standardization).

3.2 *The Ratio of Internal/External Software Spending*

Using the information on internally developed software and acquired software, the ratio of internal to external software spending can be evaluated. Histograms in Fig. 2 show the distributions of the ratio values. As manifested by the histograms, the majority of the values of the ratio of internal to external software spending lie in the range of (0;1); the median value of the ratio is approximately 0.3. In other words, for a majority of CSPs, the spending on externally acquired software exceeds considerably the spending on the internally developed software and is typically even up to ~3 times higher than internal spending.

Meanwhile, for a handful of CSPs (AT&T, BT Group, Telstra, KT Corp., and Taiwan Mobile) considered here, the spending on internal software prevails. Most of these CSPs are large incumbent telecom operators that evolved from former large state (BT, Telstra, KT) or national (AT&T) telecom operators, often monopolies. Due to their historical development, they have a highly complex unique large-scale infrastructure accumulated over decades of operations, and no independent

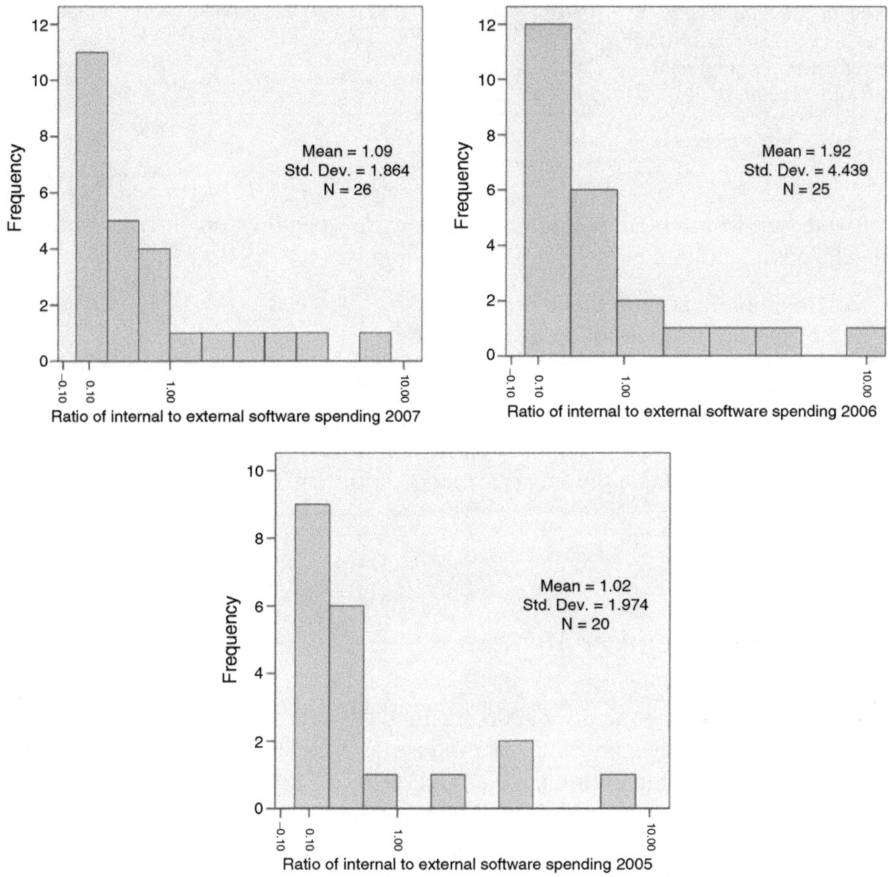

Fig. 2 Distribution of the ratio of internal to external software spending

software vendor has sufficient knowledge and experience to implement and maintain the required software.

It is noteworthy that, according to the Dittberner OSS/BSS knowledgebase, the internal software spending by the CPSs exceeds the external software acquisition. The discrepancy may be attributed to the fact that the Dittberner database better represents US-based CSPs, where internal software development is more common than in the rest of the world.

The ratio of internal/external software spending differs in different phases of software market evolution, and it can therefore be also seen as an indicator of the evolution phase. Using the Dittberner database, the relationship between the product-related and bespoken software development were approximated (see below). The product-related expenses include product license, product-related services, and integration. The bespoken software development includes only

Table 8 Share of CSP's spending on internal software development (among total software spending)

Year	Min (%)	Max (%)	Average (%)	Median (%)
2005	3.03	100.0	30.01	22.25
2006	1.23	91.16	31.54	26.68
2007	2.27	89.15	32.27	26.03

software services and integration. The share of spending on internal software development among the total software spending is:

- Approximately 80% in the first phase
- Drops to 65% and less in the second phase

For the CSPs, the estimation of the internal software spending share is presented in Table 8. As can be seen from the table, the share is in the range of 25–30%. This suggests, similarly to the software spending data, that the first phase of the evolution is over, and that the software market is currently in the Productization and Standardization phase.

3.3 *Software Delivery Methods*

Based on the Dittberner data for 2006, the total OSS/BSS revenue of these companies, excluding the revenue from other industries, was approximately USD 30,000 million (see also Table 1 in chapter, "Overview of Telecom Operator Software Market," along with related discussion). Out of this, close to USD 12,000 million was generated by software product companies from software products and related services, while the rest was generated by IT companies providing systems integration and other merchant professional IT services. Unfortunately, these figures describing market transactions exclude the in-house software development taking place within the telecom operator companies.

Two years earlier Dittberner estimated product related in-house work to be approximately 2–3 times as large as product related professional services, and the in-house bespoken work to be about 4 times as large as bespoken merchant professional services. Based on these estimates, the in-house software services accounted for 70% of the total estimated USD 49,000 million spending on OSS/ BSS software in 2004. Software licenses and application services provisioning, ASP (i.e., software "leasing" over the network) accounted together for about 7% and professional IT services (both product related and bespoken systems integration) accounted for 24% of this estimated total spending (see Fig. 3). According to the same estimates the share of product licenses, ASP and bespoken professional services are increasing while the use of in-house work is dropping below 60% of the total spending in 2008.

As shown in Fig. 3, in 2004, over 60% of spending on OSS systems was based on product-based systems implying high rate of adoption – above the critical mass.

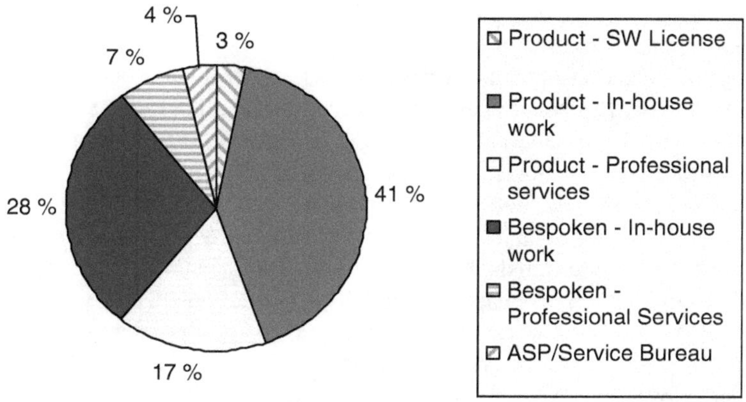

Fig. 3 Shares of software delivery methods in OSS/BSS spending in 2004 (calculated from Dittberner 2007b)

However, the available market data cannot be directly compared with the innovation diffusion theory, since we do not have figures on how large is the portion of the 650 telecom operators who have adopted OSS/BSS systems, although the general assumption is that almost all operators are also using some software for OSS/BSS purposes.

Overall, in terms of the Moore's model (cf. chapter on "Theories Related to Vertical Software Industry Evolution"), the evolution of the whole product is far from the final phase, as only 7% of the volume comes out of software licenses and ASP while most of the spending goes to services. More specifically, the market seems to be in the early majority phase of Moore's technology adoption life cycle. Nevertheless, there may also be submarkets, where services are used dominantly, and where product-based systems have not been widely adopted by the customers.

From the viewpoint of vertical software market evolution (cf. chapter on "Model for Evolution of a Vertical Software Industry"), a large portion of work – about 70% – was done in-house, although some percentage of spending goes to product licenses and ASP. This implies that the telecom software industry is roughly in the Productization phase, although some individual market segments within it may also be in other phases.

3.4 *Process, Interface, and Technology Standardization*

Let us further elaborate the analysis above on how well the OSS/BSS market matches the characteristics of the Productization and Standardization phase of the vertical software industry evolution model described in chapter on "Model for Evolution of a Vertical Software Industry."

According to the model, the core business processes of the companies operating in the market should be harmonizing in the Productization and Standardization phase. This harmonization activity is indeed taking place in the telecom industry, as the TM Forum is actively promoting the eTOM framework (TMF 2005) for defining standard processes of telecom operation.

In the Productization and Standardization phase, the software vendors are expected to strive for standardized interfaces to the underlying technologies, to eventually adopt the dominating one. In general, the OSS/BSS systems need to communicate with a variety of network technologies, such as mobile radio access networks, ATM/frame relay networks, broadband networks, transmission networks, fixed or mobile TDM, core & HLR, IP & corporate networks, Service delivery & IN machinery, as well as mobile, CPE & home terminals. However, in line with the vertical industry evolution model (see also Fig. 3 in chapter, "Model for Evolution of a Vertical Software Industry"), new software ventures would rather focus on the interfaces expected to dominate the future product markets. This hypothesis can be verified by checking if a significant correlation exists between support for a smaller number of interfaces (manifesting the focus on dominating interfaces only) and the share of product license revenue as opposed to directing software spending to in-house software development with personnel knowledgeable on the interfaces used (manifesting the young age of the vendor and hence its focus on software products). Besides the two variables above, though, it is insightful to consider also some other variables; the list of these and a brief definition for each are provided in Table 9.

The values of the Pearson correlation coefficients for some of the variables are presented in Table 10. Even with only eleven market segments, many of the indicators correlate on a statistically significant level (0.01 level marked with **) or on a statistically close to significant level (0.05 level marked with *). The variables form groups of correlating variables, which are visualized in Fig. 4.

As can be seen from the figure, the vendors operating on those market segments where the volumes of in-house software development are higher have on the average a larger size. They also support a wider range of network technologies than smaller vendors that operate in product-oriented market segments. The number of software product vendors correlated significantly with the market segment

Table 9 Variables used for analyzing correlations of technology interfaces and vertical software market evolution from internal software development to software products and services

Variable	Description
Internal-%	Share of in-house software development of the spending
Product-%	Share of software license purchases of the spending
Service-%	Share of professional IT services of the spending (assign up to 100% with the previous two)
SW vendor size	Average size of software vendor in the market segment measured in revenue
Technology coverage-%	Share of network technologies the core software product of the vendor is compatible with
Technology share-%	Market share of the technologies covered by the core product in the market segment
# of product vendors	Number of software product vendors in the market segment
Market segment volume	Total annual revenue of the market segment (M$)

Table 10 Min and max values, and the correlation matrix of the variables used

Variable	Min	Max	Internal-%	Product-%	Service-%	SW vendor size	Technology coverage-%	Technology share-%	# of vendors
Internal-%	61	71							
Product-%	4.4	12.1	−0.902**						
Service-%	24.5	26.5	−0.991**	0.834**					
SW vendor size	633	7,818	0.853**	−0.818	−0.830**				
Technology coverage-%	52	82	0.768**	−0.774	−0.735*	0.858**			
Technology share-%	59	89	0.751*	−0.785**	−0.710*	0.839**	0.991**		
# of product vendors	12	66	−0.383	0.311	0.390	−0.531	−0.063	−0.085	
Market segment volume	195	5,370	−0.356	0.183	0.397	−0.212	0.058	0.041	0.853**

In the table, the statistically significant results ($p < 0.01$) are marked with **, and close to significant results ($p < 0.05$) are marked with *

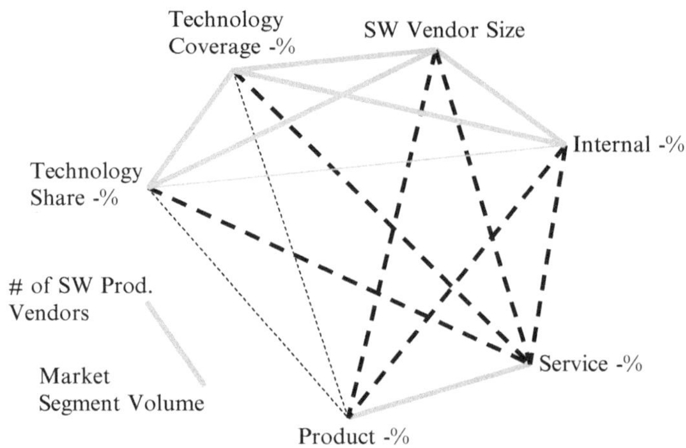

Fig. 4 Correlations of indicator variables in the OSS/BSS software market. Positive correlations are marked with *solid lines* and negative correlations with *dotted lines*. Statistically significant correlations (0.01 level) are marked with *thick lines* and close to significant correlations (0.05 level) with *thin lines*

volume, but there were no significant or even close to significant correlations between these two and other variables.

The data suggests that the share of software products used is very limited, and that a large portion of in-house software development is still taking place. This indicates strongly that the market has not reached the final phases according to any of the theoretical models presented, although the number of software vendors operating in the market seems already to be decreasing (based on Dittberner data).

The empirical market segment data supports the above hypothesis (derived from the vertical software market evolution model) that the software product vendors supporting only a limited set of network interfaces dominate the OSS/BSS market segments in the later phases. Indeed, large vendors are oriented towards IT services and have a stronger role in the market segments which are in the earlier phase of the software market evolution. These companies provide mainly professional services and have more resources and are capable to support the telecom operators who are compromised to maintain a broad variety of legacy network technologies in the earlier phases of the market evolution.

We verified, also at the level of individual companies, the negative correlation of the product orientation (the percentage of revenue coming from software product licenses) and the company size with a data set of 198 OSS software vendors. The OSS software revenue from telecom operators is negatively correlated with the share of revenue from product licenses at a statistically significant level (Spearman's rho of 0.281**). This is also in line with the whole product model, wherein the service content of the whole product decreases until the fully integrated and commoditized whole product is reached in the late majority phase.

These results suggest that the split of software spending into (1) in-house development, (2) professional IT services, and (3) software licenses can be considered as an indicator for technology diffusion and as an indicator for a phase of vertical software market evolution. It was further concluded that the market segments within a single industry could be in different phases of evolution, although they might follow a similar development trend.

4 OSS/BSS Software Market Evolution Through M&A

The operations support systems constitute a $30 billion vertical market for software vendors and system integrators. On one hand, the market is gradually maturing (see Sect. 2 in chapter on, "Analyzing the Current Phase of the OSS/BSS Software Market" and Frank and Luoma 2008). On the other, technological discontinuities bring about new opportunities and, thus, new entrants to the market. Maturity and discontinuities both in turn increase the number of mergers and acquisitions (M&A) on the vertical software market, and over 50 of M&A have occurred in the OSS market within the last seven years. M&A activity offers an interesting topic for analysis as it gives indications on the vendors' business strategies and on the future of the vertical software market.

In the following the aim is at finding out the impact of M&A activities on the structure of the telecom software market. The focus is on the horizontalization of the market that can increase as result of natural evolution of software market or through M&A activities. Horizontalization here refers to the development in software systems where is becomes possible to provide the same software system (1) for two or more customers within the same industry, (2) across technologies and (3) across industries. Horizontalization is also defined as development where a software vendor (4) expands their offering to new segments within a vertical software market or (5) enters new vertical software market.

The natural evolution enabling horizontalization occurs mainly through standardization, which reduces the barriers of entry to the software market or its segment. Standardization of business processes in a vertical industry enables providing software for many customers and reduces importance of special knowledge on vertical processes. Standard interfaces in turn enable modular systems and software product architectures and pre-integrated software products. Mergers and acquisition conversely are considered as a means to capture resources or knowledge that software vendors cannot develop internally and are unable to acquire as discrete companies. A number of contributions, in particular in the strategic management literature, links up to the motivations of M&A activities.

For achieving its aim, this section analyzes the M&A activity of the OSS market in 2001–2007, using empirical vendor data from Dittberner Inc. on 56 mergers or acquisitions. The analysis concentrates on finding the cases where merger or acquisition was committed to strengthen current market position or to enter new segments or vertical software markets. The two latter cases are seen as indicators of increasing horizontalization.

4.1 Motivations for M&A Transactions

The general rationale for mergers or acquisitions is to increase company's financial performance. Other relevant reasons include cost reductions (cf. Williamson 1975; Mueller 1980), achieving market power (cf. Stigler 1950), acquiring resources (cf. Wernerfelt 1984; Barney 1991), acquiring knowledge (cf. Grant 1996), obtaining economies of scale (cf. Stigler 1958), acquiring customer equity (Rust 2001), and managerial decisions (cf. Ansoff 1987). In the following, reasons for M&A are introduced and considered from the perspectives of both internal resources of a software vendor and competition of a vertical software market. This is elaborated through the evolution of a particular market. Consideration of the needs of companies in different lifecycle phases allows us to conclude that the grounds for M&A stressed during the market developments differ.

The vertical software industry evolution model (cf. chapter on "Model for Evolution of a Vertical Software Industry") claims that vertical software markets are born and develop according to a common pattern. The evolution of the market in this model is characterized by (a) transition from internal software development activity to outsourcing to software vendors, and by (b) increase in the level of productization. According to the model, the software development is first initiated within vertical enterprises seeking to gain competitive advantage by automating parts of their central business processes. In order to achieve this, in the initial and so-called Innovation phase, both the vertical enterprises and hardware manufacturers for that vertical industry perform internal software development activities. A major share of the total personnel is working in the secondary software industry, i.e., in vertical enterprises. In this phase, employees' knowledge on the vertical industry processes and practises as well as knowledge on the technology particular to the industry are stressed.

In the second phase of the vertical software market evolution, the competing vertical enterprises increase their software development effort and seek to copy the best practises from the others. The market leader is likely to continue its internal software development, whereas the competitors may join forces to outsource their development activities, typically to software service providers. In this productization and standardization phase, the vertical enterprises may also standardize a common technology platform or interfaces and, therefore, enable software vendors to manufacture software products cost efficiently. Also in this phase, developing software requires knowledge on processes and interface specific to the industry, although there are likely to be several different and competing micro ecologies. In this phase, software vendors which are bought typically possess the required specialized knowledge.

In the third phase, that of adoption and transition, a major increase in outsourcing of the software development activity occurs, which leads to growth of the primary

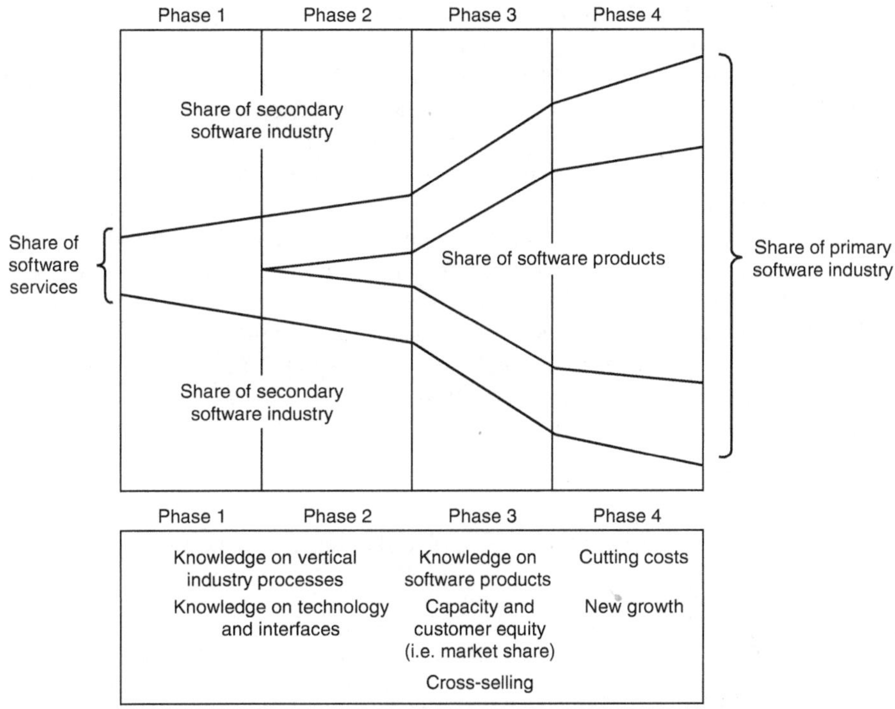

Fig. 5 The evolution of the software market (chapter, "Model for Evolution of a Vertical Software Industry") and the rationale for mergers and acquisitions in each phase of the evolution

software market and may also increase the number of software vendors in the market. This development is a result of, among other things, a further increase in the level of standardization of processes and interfaces. In this phase, to acquire specific knowledge on software products, software vendors are being bought out. On the other hand, while the software market in growing, the companies operating on it aim to obtain and establish substantial market share. Additional resources and customers are therefore also being acquired. The growth, in addition, raises the interest of companies providing horizontal software for the market. These players aim to cross-sell their products, such as databases or ERP systems, to the customers of vendors they've acquired. The vendors of horizontal software products also enter the market through acquisition to obtain the required knowledge specific to the industry.

In the last phase, before the genesis of any new market development lifecycle, one or two of the micro ecologies developed around the interfaces will have gained the position as dominant design(s). Most of the software development is related to these interfaces. In this phase, the degree of productization is high, resulting in price competition among software vendors. On the customer side, investments in software and their development decrease, which also reflects the downturn for the software vendors. Hence, it is likely that large software vendors are merged to cut high organizational overheads. At the same time, these large vendors seek growth in segments which are still growing within the vertical software market and acquire segment specific knowledge and customer equity.

It is characteristic for an OSS software system to automate one or many of the vertical telecom processes. The software is usually supported by middleware (such as database, integration and application server software), and it manages data either at the customer or telecommunication network interface. Thus, in order to enter and compete in the OSS market or its segments, a vendor must possess knowledge on vertical processes and interfaces. In Fig. 5, depicting the evolution of a vertical software market where the share of primary software industry from the total volume increases gradually, the different motivations for mergers and acquisitions are shown connected to the evolution phase. The main reasons for acquisitions in the initial two phases and in the transitional third phases are likely to be associated to acquiring resources for knowledge and capacity as well as customer relationships. In the last phase, the saturated market drives cost reductions, which are the likely reasons for mergers.

4.2 Analysis of the Impact of M&A Activities in the OSS Market

The model describing the emergence of a vertical software market also embodies the change of degree of horizontalization through standardization. In the earlier phases, the vendors providing software products are missing as the manufacturing of products for the industry is not possible. Later, standardization enables software product to appear and to be sold across industries. In the context of the telecommunications software industry, an example of the horizontalization in the earlier phase would be software vendors' ability to provide mediator software for two or more customers. Horizontalization across technologies occurred later when, e.g., the same customer relationship management (CRM) software was provided both for mobile and broadband networks. CRM software is also a good example of cross-industry horizontalization in that the same product may be provided for both telecom and financial industries by a single vendor.

In order to investigate the impact of M&A activities on the horizontalization in the telecom software market, we first assumed that company acquires resources, knowledge and customer relationships to enter the market, strengthen its position in one segment or to provide new product to a new segment within the market. These motivations are associated with the latter two phases presented above, i.e., horizontal cross-selling, market share and new growth opportunities respectively. From the assumptions we derived a classification according to which the M&A transactions can be categorized: (1) In case a vendor acquires or merges with another gaining revenue from the same segment, the transaction is labelled as *strengthening M&A*. (2) In case a vendor acquires or merges with another gaining revenue from different segment(s), the transaction is labelled as *expansion M&A* (these two cases postulate that the acquiring vendor specializes in the particular vertical software market, that is, receives over 50% of its revenue from the vertical software market). (3) In case a vendor acquires or merges with another and the acquiring company receives less than 50% of its revenue from the vertical software market, the transaction is labelled as *horizontal entry M&A*.

For the analysis a set of commercial data from Dittberner Associates Inc. was employed, consisting of a list of transactions with the vendor names of the acquirer and acquiree, combined with the total and OSS segment specific revenues of the vendors. The data set covered the years 2001–2007 for the telecom software (OSS) market, a total of 56 transactions and 236 companies. Classification of the transactions into the above presented categories was carried out by first examining the telecom software market specificity of the acquirer, to find the horizontal entry M&A transactions. Then, the transactions with telecom specific acquirer were further classified as belonging either to the category of strengthening or to the category of expansion, by comparing the revenues of the acquirer and acquiree from segments of the market. The number of cases for each year was added together to examine the trend in transactions. The results of the analysis are shown in Fig. 6.

The information of the M&A classification analysis leads us to the following conclusions. Firstly, it seems that all the hypothesized types of M&A transactions were found from the year 2001 to the year 2007. Secondly, the years 2004 and especially 2005 were active ones in terms of total number of transactions. While the revenue data also shows that the total volume of the OSS market increased from

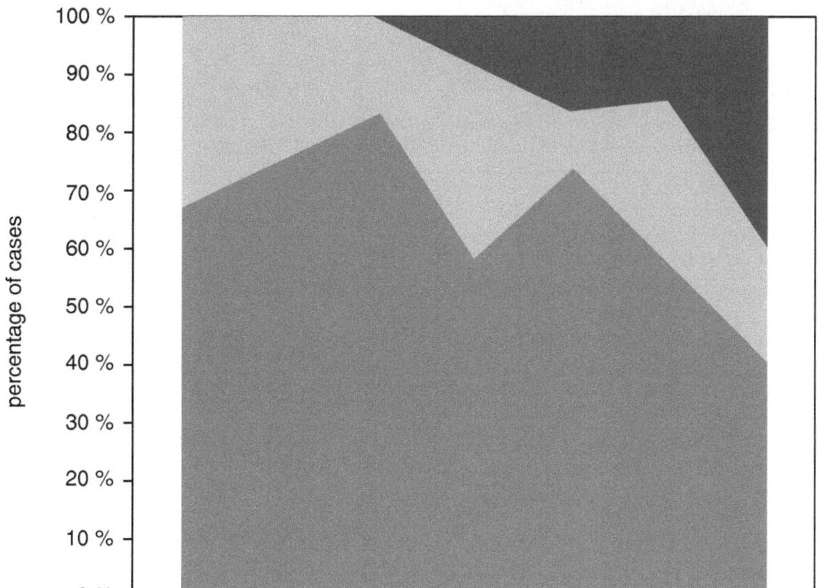

	2001	2002	2003	2004	2005	2006	2007
horizontal entry	0	0	0	1	3	1	4
expansion	1	1	1	4	2	2	2
strengthening	2	3	5	7	14	4	4

Fig. 6 The percentage and number of types of M&A transactions in the OSS market

2002 to 2005 over 30% from USD21.2 billion to USD27.6 billion (based on Dittberner data), indicating that the market was then at *the adoption and transition* phase, high M&A activity would signify a fierce battle for market shares.

Regarding the percentages of different transaction types, it seems that a major share of the transactions belongs to the strengthening category, although the trend is descending. While the share of expansion mergers and acquisitions remains approximately the same throughout the years, the horizontal entry transactions first occurred in 2004 and the percentage of that type is increasing. *This would also indicate that the software market is in the third phase of its evolution.* The entry of horizontal software providers to the OSS market is likely to have a major impact on the competition and further evolution of the market. What is interesting is that Oracle has strongly inclined towards the OSS market during the analyzed term and, having bought five OSS software vendors, gained a stable market position on several OSS market segments. The conclusion of the study, however, is that both cases, where merger or acquisition was committed to strengthen current market position and to enter new segments or vertical software markets, were found. This notion is treated as an indication of increasing horizontalization of the market.

Altogether, further analysis is still needed. The next chapter aims at clarifying the reasons of the observations by analyzing the reasons related to the special features of software business and telecom operator software market.

Part III
Issues Affecting Telecom Software Market Evolution

The previous chapter analyzed OSS/BSS software market from the viewpoints of process improvement and industry evolution models, by using both CSP interviews as well as quantitative market data. This chapter analyzes factors inherent to software and related IPR, which can, to some extent, explain the observed patterns of vertical disintegration and OSS/BSS market evolution.

Software markets generally develop from vertically integrated towards vertically disintegrated (Macher and Mowery, 2004). According to the model of vertical software market evolution described earlier in the book, innovative software is initially developed not by software vendors but in-house by companies representing the business in the market vertical. These companies invest in software development in order to automate and improve their business processes and thereby achieve a competitive advantage in a specific industry.

Vertical disintegration implies that the software is decomposed into horizontal layers; software in different layers is provided by independent software vendors and it is integrated with the other layers via standardized interfaces. Numerous vendors may operate horizontally in multiple industries, which provide a larger market for them, while some of the vendors may be specialized to serve a single industry only.

Several factors are likely to affect the vertical disintegration of software; examples of these indlude high degree of customer-specific tailoring and the need to coordinate innovation efforts spanning over several vertical layers (Mazhelis et al., 2007). In the remainder of this subsection, some of these factors are considered.

Besides prohibitively large interface implementation and maintenance efforts, vertical disintegration of a software market may be hampered by other obstacles, such as the internal complexity of the business processes being automated by the software, the need to maintain compatibility with older systems, the need to comply with the legislation mandating the use of specific systems, etc. These factors, however, have been left out of the scope of this book.

Vertical Integration due to Software Systems' Modularity

Oleksiy Mazhelis and Pasi Tyrväinen

In their product portfolio, OSS/BSS software vendors make products pre-integrated with each other as well as with other related OSS/BSS products. Prior to the full-scale development and integration, it is vital for these vendors to assess which products are reasonable to integrate. Such integration of software may be aimed at achieving different goals, such as:

- Reducing software integration efforts. Software vendors aim at maximizing their profits, whereas their customers (i.e., CSPs) aim at minimizing their costs by avoiding the integration job. As a result, to approach a win–win scenario, vendors provide already integrated products or subsystems (pre-integrated modules).
- Sharing similar knowledge base. When reusing earlier experience in implementing new products or subsystems, the vendors strengthen the quality of the implementation, as compared with the software implemented from the scratch.
- Other goals, such as sharing the same contact (e.g., technical evaluator) in the customer organization, or the like.

Besides the arguments above, integration is often used in order to facilitate innovations in software. According to previous studies on complex biological systems, improving individually one of the interconnected modules in such systems is unlikely to bring benefits at the system level. The reason for this is the propagation of the changes in the system – changes to one module bring about chains of changes in dependent modules, and the improvements in performance of one module are likely to be outweighed by decreased performance of some of the dependent modules (Altenberg 1995) which usually results in a near-zero effect on the system level. Hence, given a system of several modules, some of which are interdependent,

O. Mazhelis (✉)
Information Technology Research Institute, University of Jyväskylä, P.O. Box 35, 40014 Jyväskylä, Finland
e-mail: Oleksiy.Mazhelis@jyu.fi

P. Tyrväinen and O. Mazhelis (eds.), *Vertical Software Industry Evolution*, Contributions to Management Science,
DOI 10.1007/978-3-7908-2352-3_10,

innovations are more likely to be successful in (1) those modules that are rather independent, or in (2) an inter-dependent module if the set of interdependent modules are under the control of a single organization thereby allowing it to coordinate the innovation activities spanning several modules.

In order to study the inter-dependencies within a system, we use the so-called system modularity introduced by Frenken (2006). It indicates how independent the system's modules are from each other, and is calculated as:

$$M = 1 - \frac{P}{N} \tag{1}$$

where N is the number of modules in the system, and P is a system pleiotropy evaluated as

$$P = \log_F \prod_{n=1}^{N} P_n \tag{2}$$

Here, $P_n \in \{1, F\}$ denotes pleiotropy of module n (the number of functions affected by a change in the module); and F designates the number of functions performed by the system.

The minimum value of the modularity is obtained when each module in the system has the pleiotropy equal to the total number of functions in the system, i.e., when changes in any single module would affect all the functions implemented by the system. The maximum modularity is achieved when one of the modules has the maximum pleiotropy, while all the others have pleiotropy equal to one; the module with the maximum pleiotropy plays in this case the role of a global interface mediating the interdependencies between all the other modules.

An organization is likely to integrate closely interdependent modules into a core module, which encapsulates several dependent modules, hides the complexity of interconnections among its constituents, and provides a single interface to their functionality. After the integration, the value of the system's modularity indicator should rise, thereby reflecting the decrease of the overall system complexity.

Using the system modularity, different integration scenarios may be compared: the greater the increase in the system modularity value after the integration, the greater is the improvement in hiding complexity and creating innovation potential, and hence the more likely the integration scenario is. It is noteworthy that subsystem interconnectivity and modularity may affect the process of vertical disintegration of the software: if a vertical integration gives a better system modularity, as compared to the horizontal integration, the process of vertical disintegration may be hindered.

In order to get an insight to how the system modularity affects the OSS/BSS software, we use a TMForum's technical program, called the OSS through Java (OSS/J) initiative, which develops application programming interfaces (APIs) aligned with the TMForum's NGOSS framework. In the frame of OSS/J, a roadmap of APIs is elaborated, and for each OSS/BSS application area, a software specification, a reference implementation, and a technology compatibility kit is provided.

In the software API specification and implementation roadmap documents, the APIs to be implemented are listed, and their dependencies are illustrated (OSS/J Product Team 2006). These dependencies are used below in order to compare different scenarios of integrating OSS/BSS software subsystems. The scenarios to be compared are derived from the operations process areas specified in TMForum's eTOM. Within the operations process area, eTOM distinguishes four vertical process groupings (Fulfillment, Assurance, Billing, and Operations Support and Readiness (OS&R)) and three horizontal process groupings (Customer Relationship Management (CRM), Service Management and Operations (SM&O), and Resource Management and Operations (RM&O). Therefore, the following seven integration scenarios are compared:

- Vertical integration of Fulfillment software
- Vertical integration of Assurance software
- Vertical integration of Billing software
- Vertical integration of OSR software
- Horizontal integration of CRM software
- Horizontal integration of SM&O software
- Horizontal integration of RM&O software

These scenarios are compared based on the change in the system modularity they bring, as compared with "no integration" scenario. The resulting system modularity values are shown in Table 1. The first row in the table corresponds to the null scenario, denoted as "None," where no integration is assumed. Note that only the modules with multiple dependencies (i.e., $P_n > 2$) are integrated.

As can be seen from the table, the greatest increase in system modularity is achieved in the case of the Horizontal-CRM integration scenario. The Vertical-Assurance, Horizontal-RMO, and Vertical-Billing scenarios also provide a noticeable improvement in modularity. The Vertical-OS&R scenario and the Horizontal-SM&O scenario (which roughly equals to introduction of a SDP/SDF) introduce the least increase in modularity, while the Vertical-Fulfillment scenario gives an even lower modularity (and therefore greater complexity) than the "no integration" scenario. This suggests that the three latter scenarios are less likely,

Table 1 Comparing integration scenarios in terms of system modularity

Integration strategy	Best modularity index (only modules with $P_n > 2$ are integrated)
None	0.7058
Vertical – fulfillment	0.7007
Vertical – assurance	0.7172
Vertical – billing	0.7109
Vertical – OS&R	0.7097
Horizontal – CRM	0.7494
Horizontal – SM&O	0.7081
Horizontal – RM&O	0.7111

since they are less likely to bring benefits in innovation potential to the organizations following them.

Integrating CRM applications gave the greatest modularity improvement, due to the fact that the CRM applications are closely interrelated. These CRM applications, which include customer management, order management, problem handling, customer SLA management, and billing and collections management, have a significant number of interdependencies. Therefore, innovating in either of these systems requires an organization to have a control over the other CRM applications, thus making this integration strategy worthwhile.

The relatively small modularity improvement of the Vertical-OS&R scenario can be attributed to the high integration of the OS&R applications already in the null (no-integration) scenario. Indeed, the OSS/J roadmap standardizes inventories within horizontal layers, resulting in a single inventory per layer as opposed to a set of application-specific inventories at each layer.

Thus, in summary, the system interconnectivity and system modularity can be used in order to compare alternative scenarios of integrating (OSS/BSS) software from the perspective of the innovation potential resulting from the integration. Based on such system modularity analysis, the integration is found to be likely in the CRM and the RM&O layers of the OSS/BSS software, as well as in the assurance vertical.

Vertical Integration due to Small Market Size and High Product Development and Integration Costs

Oleksiy Mazhelis and Pasi Tyrväinen

An independent software vendor (ISV) recovers the costs of software development and configuration through license fees. However, if the number of ISV's customers is limited, then either the license fees needed for recovering the software development investments may be too high (and hence the customer would prefer to develop the software internally), or the ISV's margin may have to be decreased. Given the costs of development and configuration, as well as the margins set by the ISVs, it is possible to estimate how many customers an ISV has to have in order to recover its costs. Furthermore, given a market of a specific size and the minimum number of ISV's customers, it is possible to assess the maximum number of ISVs that can profitably compete in the market. This number may impact the evolution of the software market – if the maximum number of ISVs is small, the market may not attract enough competition, and the expected evolution of the software market may halt without proceeding to the vertically disintegrated layered set of software products. Below, analytical expressions for such estimates are provided, and are then applied to the context of the OSS/BSS software market.

1 Estimating the Maximum Number of ISVs in the Market

Let us compare the costs (1) in case the customer (the CSP) develops the software in-house, and (2) in case the software is provided by an ISV subject to a license fee.

The customer's costs in the case of buying the software include:

O. Mazhelis (✉)
Information Technology Research Institute, University of Jyväskylä, P.O. Box 35, 40014 Jyväskylä, Finland
e-mail: Oleksiy.Mazhelis@jyu.fi

P. Tyrväinen and O. Mazhelis (eds.), *Vertical Software Industry Evolution*, Contributions to Management Science,
DOI 10.1007/978-3-7908-2352-3_11,

- The costs of acquiring a license for the software product (C_{prod})
- The costs of integrating the acquired product with the other deployed systems (C_{integr})

The software license fee, on the other hand, depends on the ISV's costs which are proportional to:

- The costs of designing, implementing, testing, and configuring the initial version of the software ordered by or otherwise aimed at the first customer (C_{init}).
- The average costs of implementing added functionality demanded by each customer, including the costs of reorganizing the software architecture when necessary (C_{add}).
- The average costs of configuring the software to work in the operational environment of the customer (C_{conf}).

It is assumed that additional functionality (or interfaces to new systems) need to be implemented for each new customer, and therefore $C_{\text{add}} > 0$. Together, C_{add} and C_{conf} comprise the customer-specific costs. The above ISV costs, along with the margin, determine the price of the software license:

$$C_{\text{prod}} = \frac{C_{\text{init}} + (n-1)(C_{\text{add}} + C_{\text{conf}}) + m}{n}, \tag{1}$$

where $n$is the number of customers, and m is the cumulative margin.

Let us assume that, for the customer to opt for acquiring the software, the costs of buying and integrating a product should not exceed the costs of in-house development, i.e.:

$$C_{\text{prod}} + C_{\text{integr}} \leq C_{\text{intdev}}, \tag{2}$$

where C_{intdev} are the costs of developing and maintaining the software internally. Therefore, the sum of the initial development costs (counted on per-customer basis), the additional costs per customer, the margin (counted on per-customer basis), and the integration costs should not exceed the costs of developing software in-house, as is illustrated in Fig. 1.

Based on the above assumptions, the following boundary conditions can be derived:

$$\begin{aligned} n &\geq \frac{1 - c_1 + m/C_{\text{init}}}{1/p - c_1 - c_2}, \\ m &\leq n\big(C_{\text{init}}(1/p - c_1) - C_{\text{integr}}\big) - C_{\text{init}}(1 - c_1), \\ c_1 &\leq \frac{C_{\text{init}} + m + n\big(C_{\text{integr}} - C_{\text{init}}/p\big)}{C_{\text{init}}(1-n)}, \end{aligned} \tag{3}$$

where $c_1 = \frac{C_{\text{add}} + C_{\text{conf}}}{C_{\text{init}}}$ and $c_2 = \frac{C_{\text{integr}}}{C_{\text{init}}}$.

As can be seen, neither the absolute value of the vendor's customer-specific costs nor the absolute value of the customer's configuration costs directly affects the minimum number of customers, whereas their ratios to the initial development costs do.

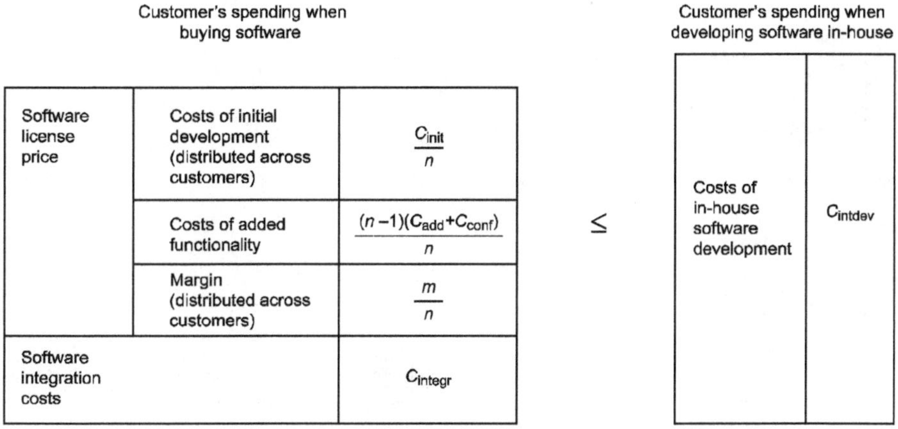

Fig. 1 Costs of buying software vs. developing software in-house

Table 1 Minimum number of ISV's customers

Independent variable	Margin (m)	Customer's integration costs (C_{integr})	Degree of ISV's customer-specific costs (c_1)
Partial derivatives	$\frac{\partial n}{\partial m} > 0;\ \frac{\partial^2 n}{\partial m^2} = 0$	$\frac{\partial n}{\partial C_{integr}} > 0;\ \frac{\partial^2 n}{\partial C_{integr}^2} > 0$	$\frac{\partial n}{\partial c_1} > 0;\ \frac{\partial^2 n}{\partial c_1^2} > 0$

Let us now consider the dependency between the minimum number of customers and the margin (m), the customer's integration costs (C_{integr}), and the degree of customer-specific costs (c_1) by the software vendor. The respective derivatives of min n are considered in Table 1; the variables were assigned the values as follows: $C_{init} = 10$, $C_{conf} = 2$, $m = 30$, and varying values were assigned to C_{add}. These dependencies are also illustrated in Fig. 2 with exemplary graphs.

As can be seen from the graphs shown in Fig. 2, the minimum number of customers grows linearly with the margin, but hyperbolically with the integration costs and the customer-specific costs. Thus, the number of customers is affected less significantly by the margin the ISV imposes, but it is highly sensitive to the costs of integration and to the additional per-customer costs. This suggests that the ISV should devote significant efforts to minimize the integration costs and the additional per-customer costs, while relatively smaller benefit would be achieved if the margin were reduced.

Finally, the estimated minimum number of customers can be used in order to estimate the maximum number of ISVs in the market as:

$$\max\ v = N/_{\min\ n}\,, \tag{4}$$

where N is the total number of customers in the market, and $\max\ v$ is the maximum number of vendors. Note, that if $N < \min\ n$, then $\max\ v < 1$, i.e., the number of customers is insufficient even for a single ISV; in this case, the software market is unlikely to emerge, and the customers will likely develop all the software internally.

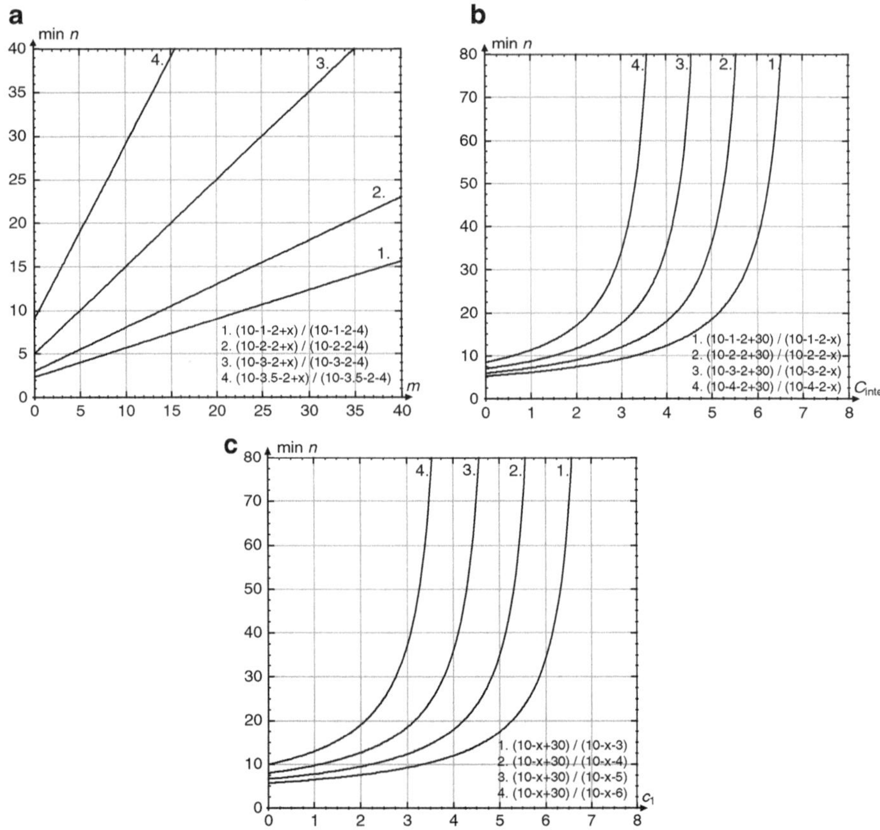

Fig. 2 Graphs depicting the dependency of the minimum number of ISV's customers on other variables: (a) margin; (b) customer's integration costs; (c) degree of ISV's customer-specific costs

In segments with (1) high software development costs, (2) heterogeneous (also) legacy systems to integrate with, and (3) small pool of potential customers, the number of software vendors is unlikely to be significant. As a result, the expected evolution of the software market may stop, without proceeding to the vertically disintegrated layered set of software products with standardized interfaces between them. In this case software will be developed in-house or delivered as be-spoken (customized) services tailored for each customer by IT service houses.

2 Example of Mediation Software

Let us apply the above model in order to assess the minimum number of customers and the maximum number of ISVs in the context of telecom billing mediation software. This software is responsible for gathering raw data describing service usage, aggregating this data and converting it into billing statistics. The raw data is collected by collectors/adapters each targeting a specific type of network elements (controllers, routers, switches, transceivers, etc.); the collected data is aggregated and processed by the mediation software, and the results are delivered to a billing system (BS). Besides BS, also fraud, revenue assurance, dataware, interconnect etc. systems may be the recipients of the information.

In order to assess the minimum number of customers needed, the ratios c_1, c_2, and m/C_{init} have been estimated by consulting the Dittberner's OSS/BSS KnowledgeBase, and by interviewing OSS/BSS domain experts. According to the information obtained, the ratios take the following values: $c_1 \approx 0.2$, $c_2 \approx 0.1$, and $m/C_{\text{init}} \approx 3$. The cost of developing for reuse is 1.5–2 times greater than the cost of developing a non-reusable equivalent (Poulin et al. 1993; Frakes and Terry 1996); therefore, p is assigned the value of 1.75. Then, the boundary condition for the number of customers can be estimated:

$$n \geq \frac{1 - 0.2 + 3}{0.57 - 0.2 - 0.1} = 14.07. \tag{5}$$

Thus, a vendor of telecom billing mediation software should have at least 14 operators as its customers. Given approximately 650 CSPs globally, the maximum number of software product vendors in this market segment is 650/14–46. This is quite close to the information provided by the Dittberner's OSS/BSS Knowledge-Base for 2006, wherein 42 software vendors offering mediation software are reported.

The business processes automated by the software in the higher layers of the OSS/BSS software (i.e., CRM and Service software layers) are closer to the end-customers and are less technology-specific or telecom-specific. These processes may have similarities with processes in other industries, and therefore, the software developed for telecom purposes may be marketed in these industries, too. As a result, the increased total pool of customers will allow a larger number of vendors to co-exist.

The ratio of the actual number of vendors in the market to the estimated maximum number ($v/\max v \in [0,\ 1]$) may be used as a measure of market concentration and competition as a software market competition index: if the ratio is close to 1, the market is considered to have a high level of competition.

Given 42 vendors acting in the mediation segment, we obtain $v/\max v = 42/46 = 0.91$. This would indicate the market being relatively competitive. A calculation of the Herfindahl–Hirschman index (HHI index) – commonly used for assessing market concentration – for the mediation segment confirms this result. The HHI for the mediation segment is 704 for the year 2006. The mediation segment can thus be concluded as being unconcentrated and competitive, and this is in line with the above calculated competitiveness share of 0.91.

Market Polarization due to Difference in Interface Implementation Efforts

Oleksiy Mazhelis and Pasi Tyrväinen

Another factor potentially affecting the evolution of the OSS/BSS software market is the high complexity of software interfaces. The software provided to a customer – a CSP – needs to be integrated with a number of heterogeneous subsystems deployed by the customer. If the number of integration interfaces is high, a vast amount of special knowledge is needed in the vendor organization. High integration efforts also take up the time of the compentent employees of the vendor organization, thereby reducing the number of customers which it is capable of serving. As a result, only few large vendors can compete in such market, and, due to a lack of competition, the evolution towards a horizontalized market with standardized interfaces and established standard architectures may be delayed or may never take place.

In this section, we shall consider whether the high complexity of the OSS/BSS software interfaces may serve as a potential hindering factor for OSS/BSS market horizontalization. OSS/BSS software provided by a software vendor often needs to be integrated with a large number of heterogeneous subsystems. If the interface implementation efforts are high, we assume that the number of companies capable of providing necessary integration decreases, and hence the OSS/BSS market horizontalization may be delayed.

An *interface* in a software system can be defined as a stack of protocols and associated data formats that govern a communication between software subsystems. Protocols in the stack may have different versions; two interfaces comprised of distinct versions of the same protocols are referred to as *variations* of the interface.

Given an interface to be supported, human resources are required not only to implement or configure the software providing the interface, but also afterwards – for maintenance, for reconfiguration of a standardized interface, or in order to

O. Mazhelis (✉)
Information Technology Research Institute, University of Jyväskylä, P.O. Box 35, 40014 Jyväskylä, Finland
e-mail: Oleksiy.Mazhelis@jyu.fi

P. Tyrväinen and O. Mazhelis (eds.), *Vertical Software Industry Evolution*, Contributions to Management Science,
DOI 10.1007/978-3-7908-2352-3_12, © Springer-Verlag Berlin Heidelberg 2009

perform new integration projects involving the interface. Therefore, human resources are needed for each interface for the entire lifetime of that interface.

Let us consider the efforts devoted by the software vendors for interface implementation activities. Due to different interfaces used, two types of CSPs are considered: incumbent CSPs and commencing operators: the incumbent CSPs are likely to have a large number of heterogeneous subsystems, often with proprietary interfaces, whereas commencing operators are expected to deal with streamlined systems having a limited set of standardized interfaces.

The assumption is that the "older" the operator, the larger the number of heterogeneous subsystems the operator has adopted and has to maintain; this heterogeneity stems, e.g., from the mergers and acquisitions, upgrades to new versions and types of equipment, etc. The incumbent operators have highly complex systems composed of a large number of diverse subsystems with complex (also proprietary) interfaces between them. New operators who may have started with a greenfield implementation, on the other hand, are likely to operate with a more manageable infrastructure with a smaller number of harmonized subsystems where standard interfaces are used more often. Interface implementation efforts differ dramatically among the two.

As a result, new entrants (software vendors), who have a rather limited number of personnel available, have a limited capability to cover integration work only in upcoming markets where the integration work is simplified by the use of web-based integration technologies (based on IETF standards) as well as by the use of business process management tools.

The total efforts for interface implementation encompass:

- Initial implementation of interfaces, i.e., initial implementation of protocol stacks
- Development of new variations of interfaces
- Configuration of the interfaces for individual customers, mainly consisting of adapting data formats to the needs of an individual customer; and
- Maintenance of interfaces

In order to assess the total interface development and maintenance efforts, we need to first determine (on a year-by-year basis):

- For each type of protocols, the number of versions/variations
- For each variation, the efforts (initial, variant development, configuration, maintenance) needed

The efforts needed for implementing an interface greatly depend on the type of protocols being used. Many types of protocols may need to be implemented, among them proprietary, OSI based (FTAM, CMISE/CMIP), CORBA, web-based (HTTP, SOAP, LDAP, RADIUS), and other standards-based protocols (FTP, GTP, MAP, etc.). Protocols may also have several versions, hence resulting in a number of coexisting interface variations.

There is a great number of interfaces present in OSS/BSS systems. For simplicity, however, only four interfaces implemented by the mediation software subsystems are considered below:

- Charging: (1) collection interface between Mediation and Network Elements (NEs), and (2) charging interface between Mediation and Billing Systems (also fraud, revenue assurance, dataware, interconnect etc. systems)
- Configuration: (1) configuration interface between Mediation and NEs; (2) activation interface between Mediation and Service Order Management System (e.g., inside CRM, Billing, sales applications)

In order to estimate the efforts, the data has been obtained through various data collection techniques including the analysis of the Dittberner Associates' OSS/BSS Knowledgebase, documentation by TM Forum, and other publicly available web-sources, and has been complemented with the data gathered through interviews. According to the data, a few dozens of the protocol types and over three hundred interface variations need to be supported by the mediation software vendor if the software is offered to incumbent operators.

Based on the available data, the interface implementation efforts (in person-years) have been estimated. The results of the estimation are shown in Fig. 1, wherein the efforts of interface implementation are shown for two cases:

- All types of interfaces are supported (dashed line).
- Only interfaces based on standard IETF protocols (RADIUS, LDAP, HTTP, SOAP, etc.) are supported (solid line).

When serving the incumbent operators, all interface types need to be implemented; the corresponding efforts by a vendor in this case are shown by the dashed line. On the other hand, to serve new operators, the vendor may have to implement only a (IETF) subset of the protocols; these efforts are shown by the solid line. The total efforts in the first case are ca. 32 times greater than the efforts in the second case.

The estimation of the interface implementation efforts can be used in order to estimate the size of the organization in both cases. If a specific portion of staff is devoted to interface implementation efforts in the company's R&D, the size of the R&D department(s) and consequently the total size of the organization can be estimated.

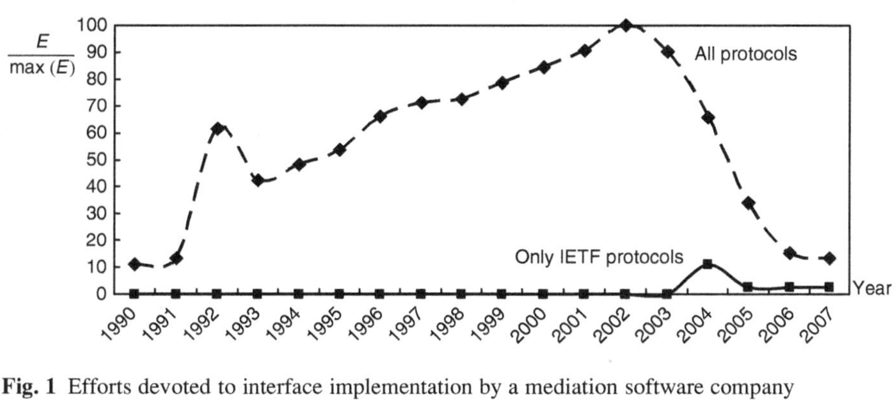

Fig. 1 Efforts devoted to interface implementation by a mediation software company

Assuming that 5–10% of personnel are dealing with interface implementation and maintenance, it is possible to infer that:

- In the first case, the organization is likely to have a few hundreds of employees.
- In the second case, the size of the organization is estimated to be a few dozens of employees (assuming that all protocols are to be developed within two years).

Therefore, only relatively large vendors are capable of serving the incumbent operators, due to large efforts required. On the other hand, it is much easier for new software vendors (especially those with highly limited resources) to serve new operators, which require only IETF interfaces to be implemented. As a result, a significant number of newcomers are likely to compete for the market of such new operators.

Thus, higher interface implementation efforts represent a threshold for the vendors entering the market of incumbent operators thereby effectively disabling the entries by small commencing vendors. Thus the high required vendors size is a real market entry barrier to new entrants in the OSS/BSS software market if also legacy network interfaces need to be supported. In a longer run, such a threshold can be assumed to reduce the number of software vendors in the market, causing a delay in the vertical disintegration of the market.

The above analysis suggests that a market polarization scenario illustrated in Fig. 2 is likely, as long as a fraction of the customers in the market require a large number of complex interfaces to be supported, while the other customers request only a small set of simpler interfaces. A new software vendor entering such

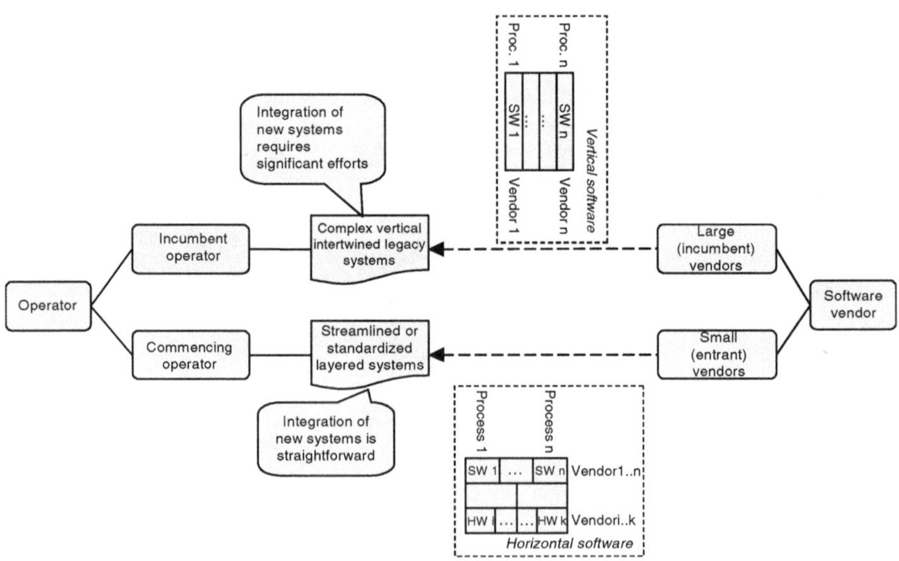

Fig. 2 Incumbent operators are served by large software vendors, whereas small vendors entering the OSS/BSS market will target commencing operators

polarized market may benefit from focusing on the submarket with the lighter interface implementation efforts. This phenomenon is also visible in the statistical analysis results represented in Fig. 4 of chap. 9. This is likely to increase the number of vendors in this new segment, thereby increasing competition which can be visible in the form of a wider variety of offering and price erosion. This chain of events is likely to benefit new operators and lower the threshold for new operators to enter the market, which will increase the volume of the market segment.

Barriers to Horizontal Development

Olli Martikainen

The growth of telecommunications has recently been based on mobile services and broadband data. According to OECD statistics (OECD 2005) the number of mobile subscribers of all telecom operators in 2003 was 828 million and in 2006 the number exceeded 2 billion. So the number of mobile connections exceeds by far the number of fixed access lines, which was 503 million in 2003. Already in 2006 the number of VoIP users exceeded 100 million, which is 20% of all fixed connections. It is expected that the Internet Protocol (IP) based traffic will soon grow tenfold compared to the amount of traditional ISDN based traffic. The transition from the existing services to Internet based ones will change the service infrastructure as well as the telecom market.

Traditional telecom and content services are vertically integrated. Each service depends on a dedicated network and corresponding terminals. Examples of such vertical services are fixed telephone services, traditional data services, mobile services and TV broadcasting services. Internet changes the vertical structure to a horizontal one, as described in Figure 1 in chapter "Analyzing the Current Phase of the OSS/BSS Software Market". All terminals and services will be Internet compatible. Instead of vertical service "pipes" there will be a horizontal structure of services, network and access (Fig. 1).

The horizontal structure will change the terminals, the services and the way the services are managed. In vertical networks the terminals are dedicated to the corresponding services. In the horizontal structure there is a freedom to combine different service functionalities in the terminal equipment. The horizontal services can be grouped to the following basic classes: Original Internet Services operated by Internet Service Providers (ISP), Application Services operated by Application

O. Martikainen

Department of Information Processing Science, University of Oulu, P.O. Box 3000, 90014, Oulu, Finland
e-mail: olli.martikainen@etla.fi

The Research Institute of the Finnish Economy (ETLA), Lönnrotinkatu 4 B, 00120, Helsinki, Finland

P. Tyrväinen and O. Mazhelis (eds.), *Vertical Software Industry Evolution*, Contributions to Management Science,
DOI 10.1007/978-3-7908-2352-3_13,

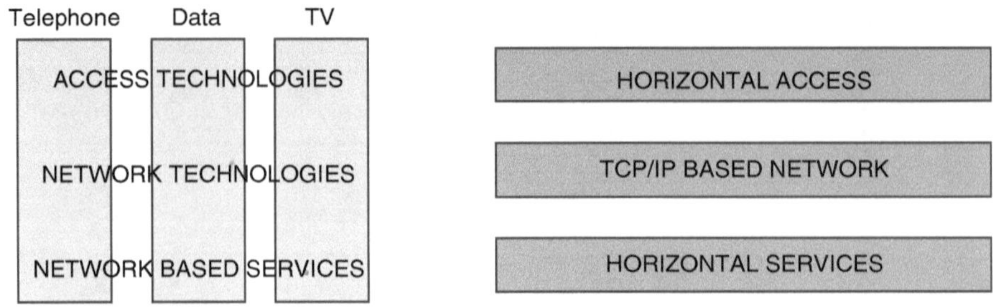

Fig. 1 Vertical services "pipes" compared to horizontal service structure

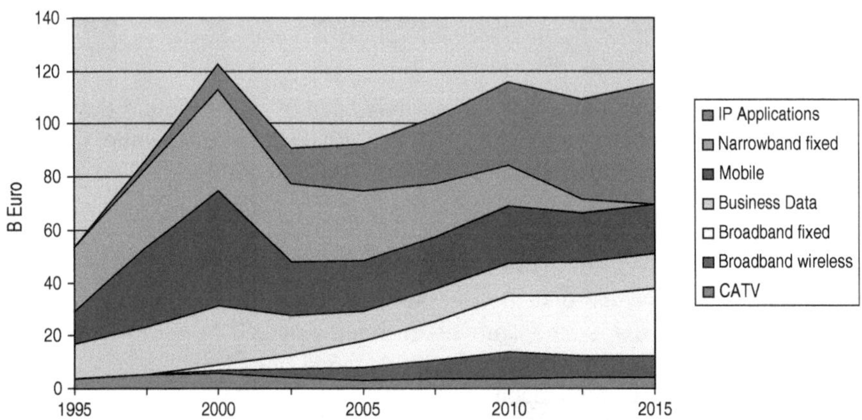

Fig. 2 European telecom investments

Service Providers (ASP), e.g., in banking and commerce, Communication Services such as Unified Messaging or VoIP by Communication Providers and Media Services by Content Providers.

In Fig. 2 based on IDC and EITO forecasts extrapolated to 2015 the IP Applications segment includes communication investments in service software for corporate, public and consumer services, which are IP-compatible, both narrow- and broadband. As we can see, the investments in narrowband fixed services, which consist mainly of telephony and ISDN technology, will end by 2015. On the other hand, the investments in broadband fixed and wireless technologies will grow substantially.

Incumbent telecom operators have market power based on network ownership especially in access networks and on service management including customer access and billing. This has been realized in GSM by the vertical integration based on the SIM-card. Incumbent telecom manufacturers have market power in their essential patent portfolios. In ETSI the patents covering some aspects of ETSI standards are called essential patents. ETSI demands that all essential patents are available for licensing on a fair basis. Major GSM manufacturers have used their GSM patent portfolios for cross-licensing purposes to obtain full patent portfolios.

Essential patent portfolios are especially large in GSM and 3G technologies (Bekkers and West 2006; ETSI 2005). The GSM-patents become obsolete when they reach the end of their lifetime after the age of 20 years, which already happened in 2008 to the first GSM-patents.

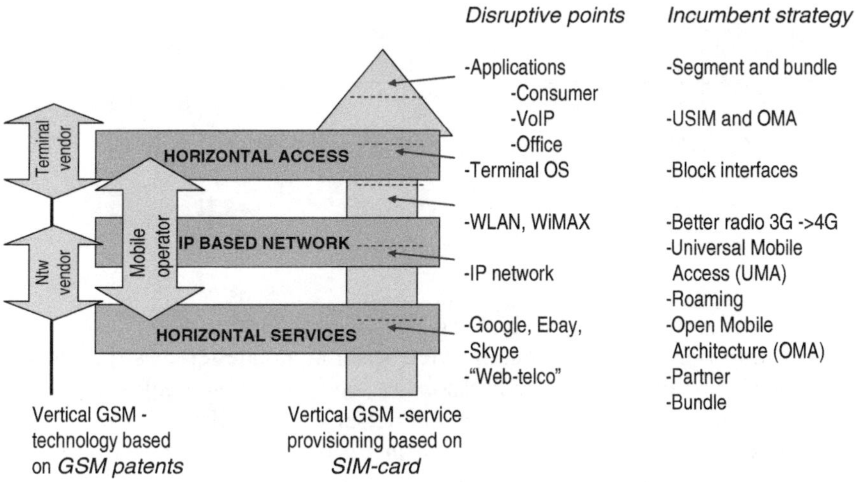

Fig. 3 Horizontal disruptive points of mobile services

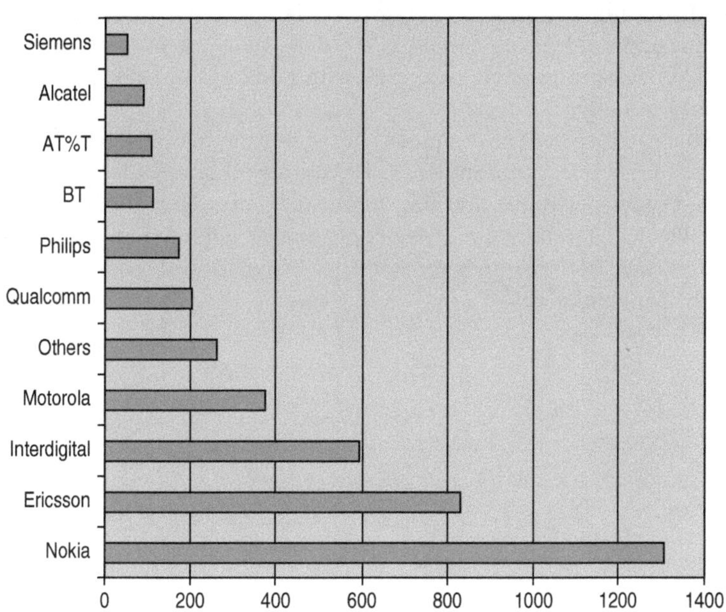

Fig. 4 Essential GSM patents

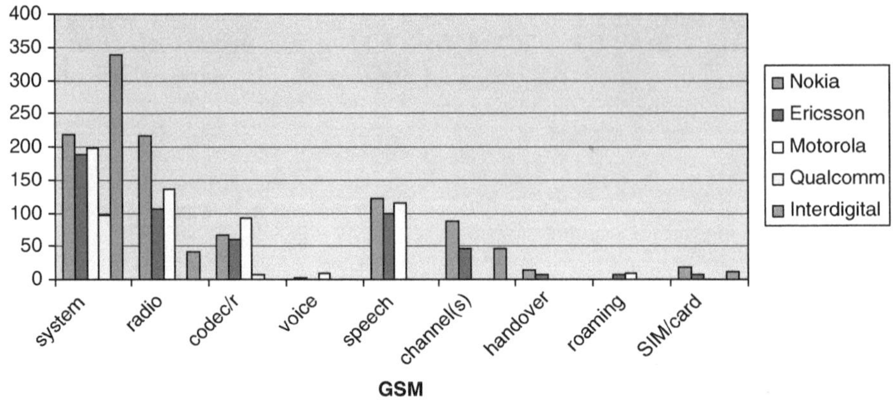

Fig. 5 Major essential GSM patent technology groups

The patent portfolios and the SIM-card based vertical integration have been the major features preventing horizontal development in mobile services. The horizontal disruptive points and the corresponding incumbent vertical strategies for preventing disruptive developments in mobile services are displayed in Fig. 3.

The essential GSM patents have been analyzed based on the ETSI essential patents database (ETSI 2005) and their total numbers with respect to patent owners are presented in Fig. 4.

The essential GSM patents were then grouped to different technology areas and the patent distribution by the number of patents is depicted in Fig. 5. From the figure we can see that the largest essential GSM patent areas are in system, radio, codec and speech processing technologies, each with hundreds of patents. The SIM-card area has only some tens of patents.

In summary, the current development in the telecommunications domain can be characterized by (1) the proliferation of IP-based traffic, and (2) gradual obsoleteness of the patent portfolios and the diminishing necessity for SIM-cards. As a result, the barriers for horizontal developments are dissolving, and, in the near future, the vertical service pipes are expected to yield to the horizontal layers of services, network and access.

Part IV
Future Scenarios for the Teleoperator Software Market and their Evaluation

The previous chapters have presented the industry evolution models available in the literature, market data analysis, results of CSP interviews, as well as the analysis of technical issues related to the evolution of the OSS/BSS market. The first section of this chapter evaluates the impact of these various factors on the future evolution of the OSS/BSS market using a Delphi study. The results are condensed into a manager's toolbox presented in the second section, which is aimed at serving as a quick set of tools for analyzing the market during the next few years. The third section describes the means to expand this toolbox. The method presented can also be used for analyzing another vertical software market as well as for creating a similar manager's toolbox for another software market.

Delphi Study

Lauri Frank

The nearby future of the OSS/BSS industry was studied utilizing a Delphi method. The Delphi method is commonly used for future studies, and it utilizes experts, who are asked to evaluate different statements or hypotheses about the future. In this section these hypotheses about the future are referred to as miniscenarios (MS). By combining several related miniscenarios, scenarios about the OSS/BSS market evolution for the next 5 years (2009–2013) can be created.

In this study, the Delphi method was employed using the following steps:

1. A set of 28 future statements to be evaluated were first generated by the research group using expert interviews, market analysis and other information as a background.
2. The set of statements was reduced to 16 most interesting statements, selected by research organization, OSS/BSS vendor and CSP representatives.
3. The likelihood and impact of the 16 future statements were evaluated on a scale from 1 to 7: For likelihood from "not likely" to "very likely", and for impact from "no impact" to "high impact". The statements were evaluated by 30 selected experts, who were asked to consider a time frame of the next 5 years (2009–2013). The expert group consisted of ten research organization representatives, ten OSS/BSS vendor representatives and ten CSP representatives. Additionally, the experts were given a possibility, by means of open ended questions, to comment on the presented statements. For the evaluation, an internet questionnaire was used. Altogether 11 experts responded to the questionnaire by the set deadline (while that number is modest, it has to be kept in mind that the quality rather than the quantity of experts counts in studies regarding future).

L. Frank
Department of Computer Science & IS, University of Jyväskylä, P.O. Box 35, 40014, Jyväskylä, Finland
e-mail: Lauri.Frank@jyu.fi

P. Tyrväinen and O. Mazhelis (eds.), *Vertical Software Industry Evolution*, Contributions to Management Science,
DOI 10.1007/978-3-7908-2352-3_14,

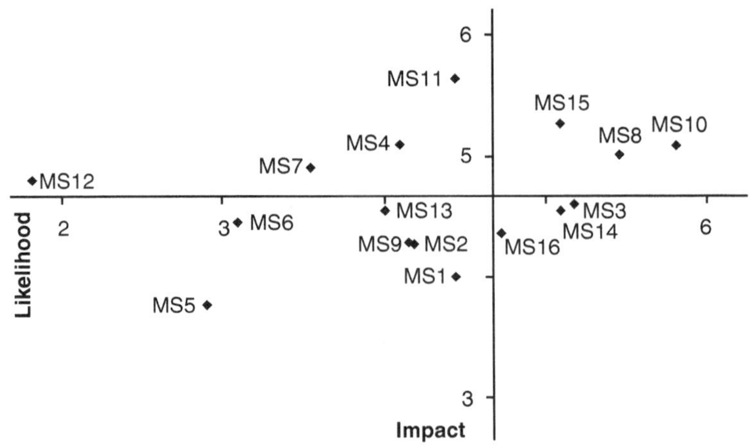

Fig. 1 The evaluated likelihood and impact of the presented miniscenarios

4. The experts then evaluated those 16 statements as regards their likelihood and impact. As a result the following four groups, based on their evaluations' average ratings in relation to the mean values, were formed:
   a. High likelihood – high impact
   b. High likelihood – low impact
   c. Low likelihood – high impact
   d. Low likelihood – low impact.

The final grouping of the 16 miniscenarios is shown in Fig. 1, followed by a discussion of their potential meaning for the future of the OSS/BSS industry.

1 High Likelihood – High Impact

These miniscenarios were rated as having a higher impact and likelihood on average. They are thus considered as likely to occur and also as having a significant impact on the OSS/BSS industry. In the following, these scenarios are described and analyzed in an ascending index order:

> MS8: *There will not be enough customers for OSS/BSS software vendors in all the product categories. Simultaneously, the software market will concentrate around software integration of merging telecom service providers rather than new product introductions.*

It is seen likely that the OSS/BSS software market will evolve more in the direction of integrating and improving existing products instead of generating and innovating new software. On one hand, this seems to be driven by the market demand: as CSP's merge with each other, they need integration of their systems. Already at the present, integration services seem to form a great part of the

OSS/BSS industry's revenue (cf. chapter on "OSS/BSS in the Vertical Software Market Evolution Model").

On the other hand, it may be thought that the existing variety of OSS/BSS software products covers all needs, and there is simply no use for any new kind of software – and thus the software development efforts will concentrate on improving existing software products. It must also be remembered that this statement was evaluated with the next 5 years in mind – in the longer run there might be room for new products also.

MS10: *The OSS/BSS software vendors expand and integrate their offering with others, either through mergers and acquisitions or through customer-specific tender coalitions.*

On one hand, this miniscenario presents a hypothesis about a consolidation of the industry. On the other hand, it is related to the aim of CSP's to be able to make their procurements from a minimum number of vendors (see chapter on "Vertical Integration due to Small Market Size and High Product Development & Integration Costs") and to increase system modularity (see chapter on "Vertical Integration due to Software Systems' Modularity"). The market analyses show that the number of vendors operating in the OSS/BSS markets has already been decreasing. The result of this statement considered as likely would anticipate a further reduction in the number of vendors in the nearby future.

MS15: *The telecom service providers are forced to adopt new organizational structures and hence also new software solutions due to the changing business environment and business models.*

On one hand, the current competitive environment puts pressures on CSP's to provide new innovative services and business models in order to be able to maintain or increase their revenue per user. On the other hand, some new services and business models might be introduced by completely new kind of CPS's, which do not yet exist on the market. This statement is seen as very likely, which may indicate that the CSP's business environment is not seen as a stagnating one, but rather as a renewing and turbulent one.

2 Low Likelihood – High Impact

This set of miniscenarios was rated as having a low likelihood but a high impact. They can thus be considered as critical for the actors' future business: They are not seen currently as a threat, but if they were to realize, they would most probably completely reshape or ruin the OSS/BSS software vendors' business.

MS4: *A pluggable service-supporting architecture with standard interfaces will be provided in a form of a single software package.*

SDP (see chapters on "Industry Structure Evolution" and "Vertical Integration due to Software Systems' Modularity") is not expected to emerge in a single product like form; instead there are likely to be various alternatives about how to organize the delivery of services. If, however, a single product would emerge, it is seen to have a high impact on the markets. This could mean that the experts see a demand for this kind of software. The demand seems to be hard to satisfy, as CSPs' existing software systems vary much, and it is hard to provide a single software which would easily fit and plug into every or most CSPs' operations.

MS7: *IT service providers (such as, e.g., Accenture) will dominate the OSS/BSS market with customized offering, based on NGOSS specifications.*

An invasion by large software vendors into the OSS/BSS markets has already started. Currently, however, these large software companies have not yet succeeded in diminishing the position of OSS/BSS specific vendors – and the experts do not see that as a likely evolution within the next 5 years. However, were they able to conquer the market, it would definitely have a big impact on the OSS/BSS industry.

MS11: *An OSS software product providing an abstraction for the IP networks, along with software products built on top of it, will dominate the OSS/BSS market after the convergence of telecom and IT.*

Currently the software products interacting with network elements commonly have proprietary interfaces. An IP based product providing an interface (standardized and uniform) to the networks would completely reshape the industry. Now the business models of many OSS/BSS vendors largely rely on the possession of information about the proprietary interfaces. If this information would become common knowledge, it could ruin their current business.

MS12: *There will be no market for commercial OSS/BSS software products as open source software is sufficient for the telecom service providers.*

Open source software is not seen currently and in the next 5 years as sufficient for all the CSPs' OSS/BSS needs. If, however, open source software systems able to reliably provide all the functionalities required by the CSP's were to exist, there would obviously be less or even no room for commercial software.

3 High Likelihood – Low Impact

This set of miniscenarios was considered as highly likely to occur in the next 5 years. Some or parts of them can already be seen as being a reality even today; however, their impact on the OSS/BSS industry remains low.

MS3: *The number of software vendors serving the telecom service providers decreases.*

This evolution has already been a reality in the OSS/BSS industry for a couple of years. The reduction in the number of vendors has partly been the result of mergers and acquisitions (see chapter on "OSS/BSS Software Market Evolution Through M&A"), and due partly to market exits. As it is already the current state of the OSS/BSS industry, it is not seen as having a high impact on it.

MS14: *The telecom service providers will increasingly outsource most of their IT operations.*

This miniscenario can also be seen as a reality of today. On the other hand, the outsourcing (see Fig. 1 in chapter, "Model for Evolution of a Vertical Software Industry") of telecom operators or CSP's does not necessarily mean a disruptive impact on the OSS/BSS industry – it can be considered as only adding on the demand and revenue of the industry.

MS16: *The telecom service providers invest on marketing, delivery and billing of media-based services.*

This statement was considered as highly likely. Already today and in the past the CSP's interest on media services has been a reality, as for example some CSP's and media houses have merged or allied together. However, considering the OSS/BSS market this evolution of some CSP's businesses does not seem to have a significant impact.

4 Low Likelihood – Low Impact

This set of miniscenarios can, in a way, be considered as the least interesting out of the 16 future statements evaluated. However, also they might contain some interesting and even critical information for the future business planning of OSS/BSS vendors and users.

MS1: *The telecom service providers will increase their use of open source software, but through outsourcing the related operations to their IT service providers.*

The use of open source is not considered as likely. One reason for this might be the uncertainties related to open source software. The CSPs often require functional reliability and guaranteed maintenance for their software systems, which open source software systems are not able to provide yet. Another reason might be that it is not seen to happen in the next 5 years, but perhaps later.

MS2: *The telecom service providers will have less bargaining power and control over the OSS/BSS software vendors due to the high switching costs.*

This miniscenario is evoked from the notion of CSPs aiming to concentrate their procurement on a limited number of software vendors. As CSPs limit their procurement and acquire large and costly systems from them, it could be hard to change the provider to a different one in the future. However, this miniscenario was evaluated as unlikely to occur in the next 5 years – and if it were to occur its impact was not considered as significant.

MS5: *There will only be a small demand for pluggable service-supporting architecture with standard interfaces.*

This statement might be looked at in reverse manner: the rated low likelihood might mean that the near future demand for SDP (see Sect. 1 on chapter "Analyzing the Current Phase of the OSS/BSS Software Market") is seen as large rather than small. Thus such architecture can be seen as welcomed in the telecom business. However, the architecture might be considered as mainly rearranging the operations of a CSP, and not so much influencing its use and procurement of OSS/BSS software. Anyhow, its impact on the OSS/BSS industry in the next following years is seen as small.

MS6: *There will be no market for independent OSS/BSS software vendors who do not ally with network element vendors.*

The experts see that at least within the next 5 years there will still be room for niche vendors, or for other vendors merging/allying with other than network element vendors. The current strategies of some network element vendors aiming to expand their operations also to other OSS/BSS software apart from network management systems is considered not to have any large impact on the OSS/BSS industry.

MS9: *New software vendors will emerge to provide software needed for the telecom service providers' new services and enhanced processes.*

This is partly a controversial miniscenario, which goes contrary to the above presented miniscenarios stating a reduction in the number of vendors. The evaluations of experts would indicate that it is not likely that new OSS/BSS vendors would emerge in the next 5 years, but that the currently existing vendors would suffice to take care of the needs of CSP's.

MS13 (OSSSW5): *There will be less demand for traditional OSS/BSS software as similar software is already embedded into network elements.*

This miniscenario of embedded OSS/BSS software is not seen as likely to occur in the next 5 years. If it were to occur, however, it is not expected to have a large impact on the OSS/BSS industry – maybe because also the embedded OSS/BSS software needs to be manufactured by some software vendor.

5 Other Miniscenarios with Potentially High Impact on the Telecommunications Software Market

Twelve further miniscenarios were found potentially interesting in terms of the impact they may have on the telecommunications software market. Based on discussions with domain experts, it was decided to leave these outside of the scope of further consideration. Therefore, these miniscenarios are listed below only for reference. However, some of these miniscenarios may also provoke fruitful ideas for new innovations and distractive market evolution scenarios.

1. The software procurement of a new, commencing operator better enables the use of standard OSS/BSS products.

2. There will be no single solution for providing telecommunications services in the future.
3. General software companies will expand to the resource management layer, and the current competitive advantage of the network equipment vendors will diminish.
4. The process of market consolidation will continue within the regulatory limits.
5. The threshold for the entry of new software vendors will increase.
6. The focus of software vendors will shift away from OSS/BSS, and OSS/BSS software will become more standardized.
7. There will be no unified dominating network resource management software.
8. Software integration will have an increasingly important role in the OSS/BSS market.
9. There will be no dominant / unified standard for OSS/BSS software interfaces.
10. Operators with processes fully automated with software will gain a greater market share.
11. Telecom operators will direct investments towards customer demands providing new opportunities for innovative software at the customer interface.
12. Regional operators profile themselves as ICT-service providers, focusing on serving local SMEs and public sector organizations by providing both communication and IT-services.

Manager's Toolbox for OSS/BSS Market Analysis

Pasi Tyrväinen

The purpose of this section is to provide a set of quick tools for analysts and managers to analyze the status of the telecom operator software market.

1 Base-line Scenario for OSS/BSS Evolution

The base-line scenario in Fig. 1 represents the expected evolution of the global telecom operator software market. It is based on matching the contemporary OSS/BSS market data and past events to the generic evolution model of a vertical software industry. The figure includes layers for telecom operator business and telecom operator processes at the top and a layer for technology and technology interfaces at the bottom. These three layers have an impact on the OSS/BSS software market (in the middle) and on the emergence of software vendors to these markets.

The market evolution, represented along a time-line from left to right in the figure, starts with national and regional telecom operators implementing in-house software systems and network equipment vendors implementing vendor-specific software for their systems. The key important events in the past shown in the figure include the introduction of early ITU standards, the introduction of mobile communication technology and IP technology, as well as the deregulation of telecom business.

The current situation is presented in the middle with an ellipse-shaped shadow, and the key elements of alternative future miniscenarios are shown on the right hand side. At present, IP technology has been widely adopted, but the all-IP approach is not yet dominating due to several reasons mostly associated with legacy

P. Tyrväinen
Department of Computer Science & IS, University of Jyväskylä, P.O. Box 35, 40014, Jyväskylä, Finland
e-mail: Pasi.Tyrvainen@jyu.fi

P. Tyrväinen and O. Mazhelis (eds.), *Vertical Software Industry Evolution*, Contributions to Management Science,
DOI 10.1007/978-3-7908-2352-3_15,

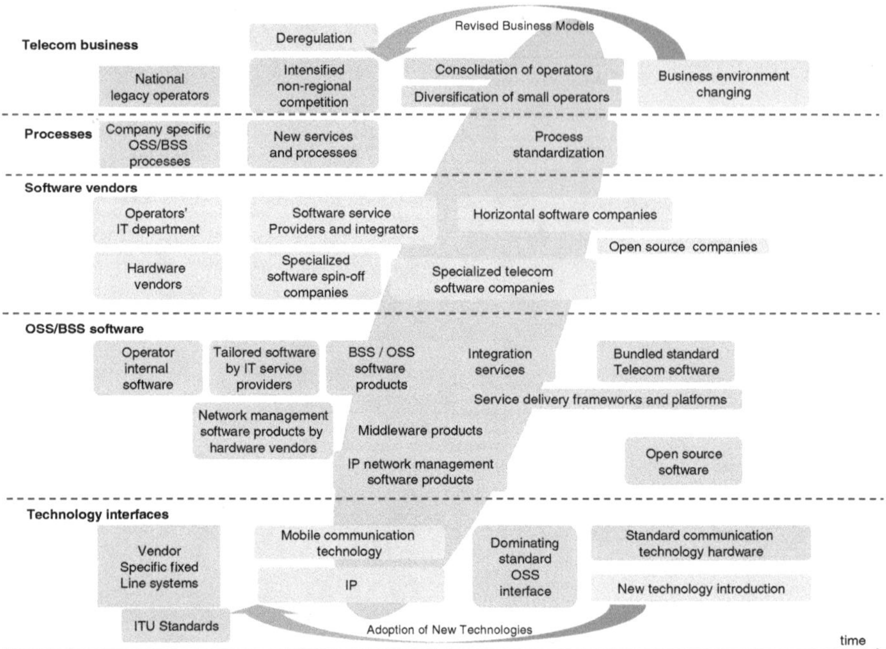

Fig. 1 The base-line scenario for OSS/BSS software market. The current situation is indicated with an *ellipse-shaped shadow*

installed base (high volume of capital expenditure embedded in equipment to be demortalized, costs in upgrading and harmonizing heterogeneous technologies and systems etc.) and due to immaturity of replacing technologies and business models. This is visible on the CSP business side as the large legacy operators use in-house resources and external software services with wide competency portfolio to maintain their legacy systems. Large IT service houses try to support this with TM Forum standardizing OSS/BSS systems. New operators are, in general, more willing and capable to adopt software products based on all-IP approach, but have to rely on their limited resources in building their own infrastructure or on regulation providing space for NVOs. Over half of new software system implementations are based on software products, but most of OSS/BSS software spending is still on IT services and the share of commercial off-the-self licenses is still a small percentage of total SW spending.

Assuming the evolution of the OSS/BSS software market to follow the generic vertical industry software market model, the next events to take place are adoption of a dominating standard for the technology interfaces and standardization of OSS/BSS business processes enabling implementation of less complex software systems in the form of cost-efficient and effective products. If this does not happen, the software development activity and business will find new growing markets to make best commercial use of the software engineering resources and the OSS/BSS software market will continue its consolidation.

Alternatively, the continuous introduction of new technologies can prevent the adoption of dominant standards and the market will remain in the current phase of competing standards. Further, new business models adopting new combinations of traditional telecom operator processes and processes of other vertical industries (e.g., media, finance, IT services) may create new needs for new software solutions starting a new evolution cycle for emerging OSS/BSS industry subsegments.

2 Miniscenarios

The following miniscenarios represent specific events that either freeze further evolution or bring it a step further or have other major impact to the market evolution in the time frame of 2009–2013. The likelihood and impact values are presented for those mini scenarios evaluated in the Delphi study (see Chapter 14). The miniscenarios are positioned into layers, actors, and activities in Fig. 2.

2.1 *High Likelihood: High Impact Miniscenarios*

These three miniscenarios represent evolutionary development of the OSS/BSS market with likelihood and impact higher than the average of the miniscenarios.

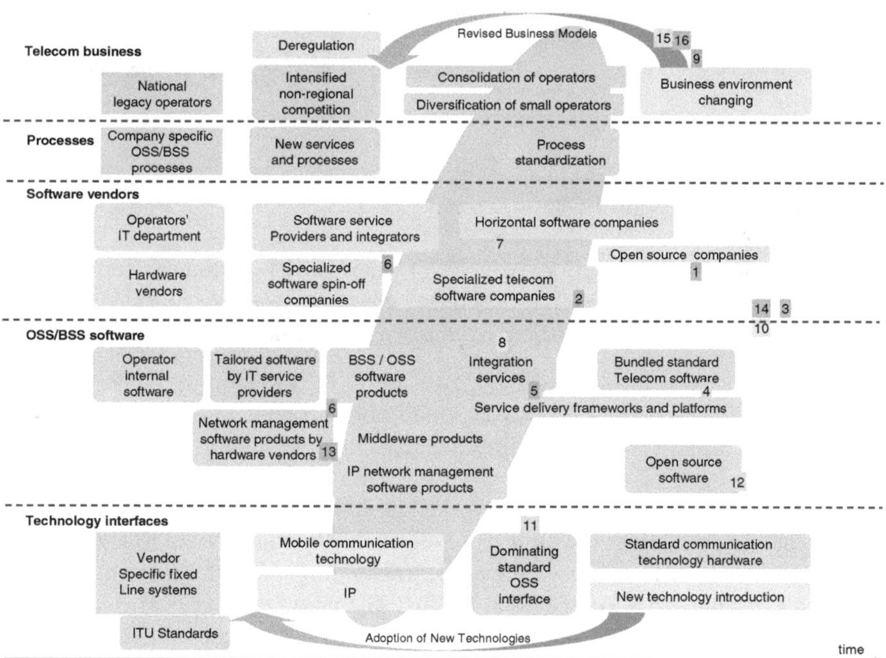

Fig. 2 Miniscenarios positioned into the base-line scenario

All actors in the market are expected to plan for these, i.e., (1) for maturing product markets, where small software products and vendors are integrated to large offerings and large vendors, (2) for an increased need for integrating heterogeneous legacy software of consolidating telecom operators, and (3) for an emergence of new software products for supporting new business models and processes of CSPs.

2.1.1 MS10: Integrated Offerings with Fewer Vendors

Description: The OSS/BSS software vendors expand and integrate their offering with others, either through mergers and acquisitions or through customer-specific tendering coalitions.

Reasoning: CSPs wish to minimize the number of software vendors for OSS/BSS software, which will also reduce transaction costs on software procurement. This is in line with consolidation of the OSS/BSS software industry as the growth rate of the market is modest. The market analyses show that the number of vendors operating in the OSS/BSS markets has already decreased approximately 14% during 2002–2006.

Implications: Software vendors should find a position and/or alliance for themselves in a market with a half of the vendors left.

Metrics:

- Number of software vendors reducing gradually to about a half of the current number (i.e., 150 or less).
- Number of software vendors can remain in a higher level if market growth exceeds 10%.

2.1.2 MS8: Integration Focus

Description: CSP consolidation reduces the number of customers for software products and requires services for integrating the systems of merging CSPs (during 2009–2013). The emphasis is shifted from software products to integration services.

Reasoning: First, in some product categories the OSS/BSS software vendors would need more potential customers (CSPs) than are currently available, in order to operate a product-based business. Simultaneously, consolidation of their customers shifts their business priority to integrating existing software systems of merging operators rather than innovating new software. Already at the present, integration services of legacy systems developed for legacy technologies seem to form a great part of the OSS/BSS industry's revenue (according to Dittberner, the system integrators receive 55% of the total OSS/BSS market volume). The variety of existing OSS/BSS software products seems also cover the needs of CSPs. There is little use for new kind of software, and the software development efforts will concentrate on improving existing software products.

Implications: Consolidating CSPs implies consolidation need for OSS/BSS software product vendors but brings (temporarily) new business for systems integrators.

Metrics:

- Number of CSPs (now round 650 mobile operators, decreasing).
- Ratio: the number of CSPs divided by the number of OSS/BSS software vendors. Threshold number of customers needed for each software product vendor in a market segment (round 14 in the example, see Chapter 11).
- Sum of market share of software integrators in the OSS/BSS market out of the total OSS/BSS software market (revenues of the software integrators represented 55% of the OSS/BSS software market in 2006, see Chapter 5).

2.1.3 MS15: New CSP Business Demands New Software

Description: Changing business environment and business models force telecom service providers to adopt new organizational structures and hence also new software solutions.

Reasoning: The current turbulent competitive environment puts pressures on CSP's to provide new innovative services and business models in order to be able to maintain or increase their revenue per user. On the other hand, some new services and business models might be introduced by completely new kind of CPS's, which do not yet exist on the market.

Implications: OSS/BSS software vendors or new entrants will appear in the market to serve new operators (and new businesses within old CSPs) with new (typically product-based) solutions. This will be an opportunity for established players to acquire new business and growth, but may also require fast creation of additional features to existing products and systems.

Metrics:

- Number of new CSPs.
- Number of new CSP service introductions.
- Number of CSP reorganizations announced.

2.2 Low Likelihood: High Impact Miniscenarios

The following four miniscenarios (by decreasing likelihood and impact) have all lower likelihood but higher impact than the average. They represent a somewhat revolutionary development for the next phase of the OSS/BSS market where either a dominant design or a dominant approach for OSS/BSS software has been adopted. This will cause a major reshaping of CSP software architecture, will revise the set of required software products, and will renew the positioning of software companies in the market.

2.2.1 MS11: IP Platform Product Establishes a Dominant Design

Description: An OSS software product provides an abstraction for the IP networks and the software products built on top of it will dominate the OSS/BSS market after the convergence of telecom and IT.

Reasoning: Currently OSS/BSS software rely on the use of network element management systems having proprietary interfaces. The proprietary information of these interfaces creates an entry barrier to OSS/BSS software developers. An IP based product providing an open interface to the networks, while hiding the variation of the networks from the OSS/BSS software to a sufficient extent, would remove this obstacle and remove this entry barrier and completely reshape the industry.

Implications: For the company providing the interface it would create a (probably nonsustainable) competitive advantage. On top of such interface the competition would be greatly intensified, generating a burst of new products. An interface covering only a small part of network equipment or an insufficient set of its features will not produce the effects fully.

Metrics:

- Market share of network equipment vendors whose (sufficient) functionality is covered by the platform.
- Market share of applications with covered functionality of the supported equipment vendors >50%.

2.2.2 MS4: Service Delivery Platform Realized

Description: A pluggable service-supporting architecture with standard interfaces will be provided in a form of a single software package.

Reasoning: Experts agree on a need for a service delivery platform bringing together interfaces needed for fast creation and provisioning of new services. This demand is difficult to satisfy due to variation in CSPs' legacy software systems and difficulties to fit in a single SDP into most of CSPs' operations. A common belief is that there will not be a single SDP product but rather various alternatives to organize service delivery. If, however, a single product were to emerge, it would have a high impact on the markets.

Implications: An SDP with a major market share would form a dominant design attracting software development activity around compatible pluggable components. This ecosystem would outperform competition in the breadth of offering, cost, availability of support services, and size of customer base. Independent software vendors would have to consider joining it or a comparable alliance.

Metrics:

- Share of typical OSS/BSS functionality covered by an SDP.
- Reduction of effort needed to implement a new service with an SDP (order of magnitude reduction for 80% of new services).
- Share of CSPs whose requirements an SDP product covers (reaching 15% = 100).
- Number of customers of an SDP out of CSPs (reaching critical mass with 15% of CSPs equaling to about 100 CSPs).

2.2.3 MS7: IT Service Providers Dominate with NGOSS Customizations

Description: IT service providers (such as, e.g., Accenture) will dominate the OSS/BSS market with customized offering, based on NGOSS specifications.

Reasoning: An invasion by large software vendors into the OSS/BSS markets has already started. So far these large software companies have not succeeded in diminishing the position of OSS/BSS specific vendors and the experts do not see this as a likely evolution within the next five years. This seems to be mainly due to limited domain knowledge, although they have a large pool of engineers with knowledge of a large number of network technologies.

Implications: If IT service providers were able to standardize the OSS/BSS processes they would have better competitive position due to their capability to balance the engineering load across large resource pools that have a big impact on the OSS/BSS industry.

Metrics:

- Market share of IT service houses vs. market share of product-based ISVs on OSS/BSS market.

2.2.4 MS12: Open Source Software Dominance

Description: Open source software is sufficient for the telecom service providers, leaving no market for commercial OSS/BSS software products.

Reasoning: Open source software is currently and in the next five years seen as not sufficient for all the OSS/BSS needs by a CSP. If, however, open source software systems were to exist and able to reliably provide all the functionalities required by the CSP's, there would obviously be less or even no room for further commercial software.

Implications: OSS/BSS software vendors would have to change their tactics with respect to open source software. E.g., they would have to provide their software for free (still maintaining customer lock-in), release their product as an open source software, or find some intermediate way to integrate with the open source.

Metrics:

- Market share of IT services related to open source software out of the total volume of IT services related to OSS/BSS.

2.3 Low Impact Miniscenarios

The following three miniscenarios are likely to happen unless some of the high impact miniscenarios, by introducing a dominant design, will turn the development to another direction.

2.3.1 MS3: OSS/BSS SW Vendor Consolidation Continues

The number of software vendors serving the telecom service providers decreases.

2.3.2 MS14: Outsourcing Continues

The telecom service providers will increasingly outsource most of their IT operations.

2.3.3 MS16: CSPs Focus on Customer Experience

The telecom service providers invest on marketing, delivery and billing of media-based services.

Out of the full list of miniscenarios generated for the OSS/BSS market 16 were considered as the most important ones and were included in the Delphi study. The following miniscenarios were also included in this filtered set of 16 miniscenarios, although their likelihood and impact was below the average of the top 16 miniscenarios. These miniscenarios were (in decreasing order of likelihood):

- MS1: Open source will be used through outsourcing
- MS2: High SW vendor switching costs dominate
- MS9: New vendors provide SW for new processes
- MS13: OSS/BSS software will become embedded to equipment
- MS6: OSS/BSS SW vendors have to ally with equipment vendors
- MS5: No need for SDP architectures

Developing Toolboxes

Pasi Tyrväinen

The manager's Toolbox for OSS/BSS software market analysis is being developed for the analysis of the market during the SmarTop project, and is based on the data sources described in Section "Data sources used". The method for analysing a vertical software industry which has been elaborated and used in this book is described shortly below; this method can be applied for the analysis of other vertical software markets as well as for the creation and expansion of manager's toolboxes similar to the one in Chapter 15.

The vertical software industry analysis method used has the following main phases:

- Baseline creation
- Indicator calculation
- Analysis
- Scenario creation
- Scenario evaluation

Figure 1 visualizes the first three phases related to data collection and analysis while the two latter phases are related to formalizing the results as scenarios. The five main phases are described next.

1 Baseline Creation

This phase includes identifying data sources available. Typically any vertical market is analyzed by some commercial market analysis companies, which collect quantitative market and purchase data, as well as qualitative data describing

P. Tyrväinen
Department of Computer Science & IS, University of Jyväskylä, P.O. Box 35, 40014, Finland
e-mail: Pasi.Tyrvainen@jyu.fi

P. Tyrväinen and O. Mazhelis (eds.), *Vertical Software Industry Evolution*, Contributions to Management Science,
DOI 10.1007/978-3-7908-2352-3_16,

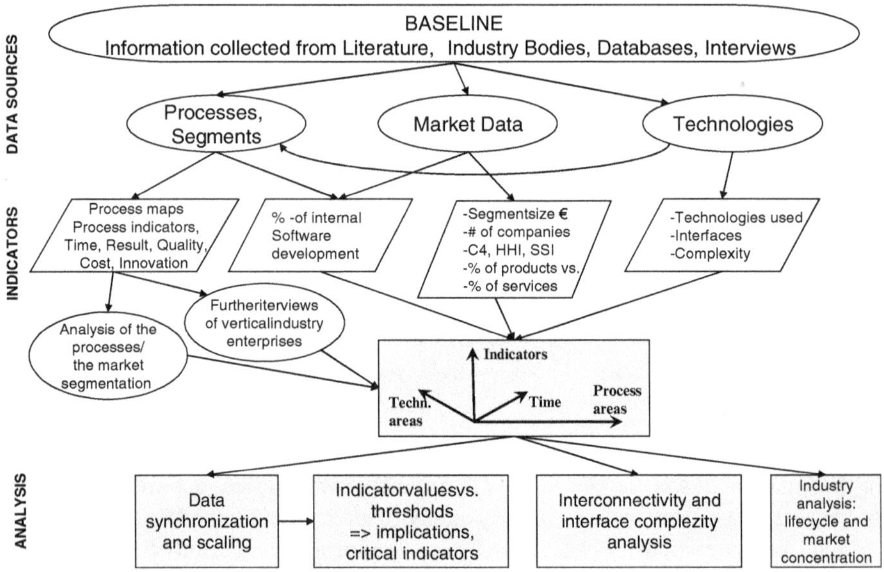

Fig. 1 Identifying available data sources and indicators and analyzing the data

technology development, evolving customer needs, and overall market evolution. In addition to the market being analyzed (e.g., software for telecom operators), the internal software development of the customers (e.g., telecom operators) and the market of the customers (telecom services provided by users of telecom software to their customers) need to be analyzed. The main areas for data collection are:

- Standards for processes of the target vertical industry (such as eTOM for teleoperator processes).
- Market segmentations of the software market analyzed (such as TMF application map for OSS/BSS software).
- Technologies and interface standards needed by the software of the vertical industry (such as communication network technologies and OSS/J in OSS/BSS)
- Software market data, including volumes of segments divided according to technologies, vendors, and delivery method (SW products, IT services, SaaS).
- Customer market data including the split of software spending into in-house development, IT services, and software product purchases.

2 Indicator Calculation

As all indicator data is typically not readily available, some data may have to be acquired with interviews, collected from various sources (e.g., Internet and annual reports) and calculated based on synchronizing and combining multiple data sources.

3 Analysis

The basic analysis of contemporary situation is based on matching the indicator data against existing industry evolution models. The basic analysis can be made by calculating the share of in-house software development, use of external IT services and purchase of SW products and matching the data with the model of evolution of a vertical industry software market. The attractiveness of the market for the software vendors can be analyzed by counting the number of vendors and the HHI index and comparing that with the industry evolution model. A detailed analysis can be carried out by applying the other methods and models presented in this book.

4 Scenario Creation

Draft scenarios for further market evolution can be created with three principal means. First, each type of the analysis above (positioning the market into some evolutionary phase) also predicts one or more likely future scenarios.

Secondly, industry experts (e.g., OSS/BSS) and target industry experts (e.g., telecom operators) can be interviewed for their view of the market evolution.

Thirdly, a systematic creation of scenarios can be followed in line with the process for developing scenarios defined by Schoemaker (1995):

1. Define the scope: time frame, products, markets, geographic areas, technologies.
2. Identify major stakeholders: customers, suppliers, competitors, government, etc.
3. Identify basic trends: political, economic, societal, technological, legal, industry.
4. Identify key uncertainties: political, economic, societal, technological, legal, industry factors.
5. Construct initial scenario themes: extreme worlds, crossing uncertainties.
6. Check for consistency and plausibility: timeframe, contradictions in uncertainties (full employment and zero inflation), stakeholders' will.
7. Develop learning scenarios: deriving more general and relevant themes.
8. Identify research needs: blind spots in uncertainties and trends.
9. Develop quantitative models.
10. Evolve towards decision scenarios: iterate 1 through 9.

5 Scenario Evaluation

Scenarios produced can be evaluated in line with scenario work as described in Chapter 14. The scenarios having a high likelihood and high impact should already have been prepared for, whereas the scenarios with high impact and small likelihood are typically the most demanding and require intensive effort for strategy creation.

Following the vertical software industry analysis method presented here will help enterprises in creating a base-line for their strategy. Periodical efforts are also needed for evaluating the likelihood of the alternative scenarios based on the threshold values defined for the indicators as well as for updating the scenarios. This method is also useful when developing and extending a manager's toolbox, i.e.,

- For adding new miniscenarios for the toolbox as described in Chapter 15 at a later date.
- For the creation of a manager's toolbox for analyzing another vertical software market.

Part V
Evolution of Telecom Operator Software Industry – Conclusions and Future

Evolution of Telecom Operator Software Industry: Conclusions and Future

Pasi Tyrväinen

1 Vertical Software Industry Analysis

Analysis of software industry is a challenging task due to the inherent characteristics of software industry. This is not simply due to the immaterial nature of software, or to zero marginal cost of replicating software, nor it is due to knowledge intensiveness of the industry. Most software is vertical software and is developed for the purposes of some vertical industry, such as finance or telecom, whose business the software will support. The major challenge for analysing a vertical software industry comes from understanding the interaction of software technology and the business requirements of the vertical industry.

If the vertical industry is mature and includes a relatively high number of enterprises with standard business processes then software can be seen as a commodity whose cost should be minimized. In this case the vertical software industry is also likely to have entered a mature stage in the industry evolution model. In line with this development, the vertical software industry has most likely reached a dominant design indicating the later phases in the vertical software market evolution model, where a few dominating companies are servicing the vertical industry with their product-based offerings. The other extreme can be found from emerging industries, which use in-house software development or IT consults to create software systems that provide competitive advantage for the enterprises creating a new industry. In both of these cases the model for industry evolution stages and the phases in the model for vertical software industry evolution go hand in hand.

However, if the number of vertical industry enterprises is too small or the technologies they need have nonstandard software interfaces, the vertical software industry evolution may freeze to the second phase, although the number of software

P. Tyrväinen
Department of Computer Science & IS, University of Jyväskylä, P.O. Box 35, 40014, Jyväskylä, Finland
e-mail: Pasi.Tyrvainen@jyu.fi

P. Tyrväinen and O. Mazhelis (eds.), *Vertical Software Industry Evolution*, Contributions to Management Science,
DOI 10.1007/978-3-7908-2352-3_17,

companies is decreasing in some market segments implying that the market has reached a mature stage in the industry evolution model. This means that the software development activity is redirected, to some extent, to some other industries although no productized solutions have been fully adopted. This also is an opportunity for service providers to offer alternative approaches to the operators, such as business process outsourcing as a mean to gain some cost advantage although not comparable to automating the processes with software.

This book describes the 2-year research effort to analyze the vertical software industry of telecom operator software, that is, the OSS/BSS software. After the theory background in Part I, Part II used multiple means to analyze the current status of the industry. The results imply that only the very largest operators still develop software in-house as a means to gain competitive advantage. However, only a small percentage of their software spending goes to software products while a large volume of effort is still directed to external software consultancy based on customer-defined specifications as well as tailoring based on core software product from an independent software vendor. (The reasons for these are elaborated in the next subsection.)

There exists also some variation between different geographical regions as well as between different market segments within the same market. In practice, the difference between market segments means that small operators have to make compromises in their IT strategy. If they are willing to adopt product-based offering available for some of their core processes and outsource their in-house software development activity, they will have to use external IT consultants for tailoring customized software for the other processes, to which there exists no suitable software products. This development explains the high share of external IT spending even if there are market segments with a reasonably high volume of product-based business. Altogether, this further implies that using a vertical software market and the enterprises within it as the units of analysis is essential for understanding a vertical software industry.

Part III analyzed some of the reasons which are likely to direct or hinder the further evolution of the vertical software industry (OSS/BSS) causing disconnection with the vertical industry of telecom operators. These include the analysis of system modularity stating that the systems are integrated along the main business processes handling the bulk of the information processing volume. The minimum number of operators as the customers for OSS/BSS software vendors was also estimated and verified with market data. Most importantly, the continuous evolution of telecom technologies seems to be a major reason for a need to renew software interfaces at a pace not possible for small software companies. In addition, IPR strategies and SIM-locking can slow down the evolution of the software market.

Part IV presented a Delphi analysis, which was used for validating alternative future miniscenarios for the OSS/BSS market. The results were summarized as a Manager's Toolbox in Chapter 15 for business users. Chapter 16 outlines a method for repeating this research in some other vertical industry as well as for extending the Manager's Toolbox presented or creating a similar Toolbox for another vertical software industry.

2 Telecom Operator Market

The telecom operator software market (OSS/BSS market) was analyzed here as an example of a vertical software industry. This analysis included the viewpoints of the business processes reflecting the telecom business, technology viewpoint visible in OSS/BSS interfaces, and market viewpoint as presented in market statistics. The results have been condensed into a default scenario representing the main past events impacting the OSS/BSS markets and the expected future predicted by the model of a vertical software industry evolution.

The OSS/BSS market is exceptional when compared to other markets, as a major attempt to standardize the business processes is underway. The efforts of TM Forum on standardizing the OSS/BSS processes with the eTOM model as well as efforts to standardize a common data model and a common set of interfaces for applications have been received well and adopted by many telecom operators and especially by IT integrators.

Software related to network interfaces has traditionally been equipment and vendor specific. Later, network technology standardization has been very active and new standards have been created at a fast pace. However, the pace of introducing new network technology providing high added value compared to previous generations has been even faster. And each new network technology with different performance characteristics has required new software. Thus the number of network interfaces to be integrated to OSS/BSS software has been overwhelming, thus prohibiting OSS/BSS software from reaching a steady market with a single dominating design.

The telecom operators having invested heavily into previous generations of network infrastructure are forced to maintain the legacy systems till the end of their investment demortalization periods. But they are also forced to provide new services enabled by the new technology as well as to provide the same services to the existing customer base still connected with the old network technology. Whenever the independent software vendors have not been able to provide software to support their set of legacy technologies, CSPs have found themselves forced to use services of large IT integrators with knowledge of a large set of technologies. This is visible in market statistics, which show that only 7% of CSPs' software spending goes to product licenses and ASP while in-house development and IT services take the rest. Also the tendency of large CSPs to use more in-house development and services of large IT service houses is much higher compared to small CSPs, which prefer the use of products from smaller vendors with associated services.

There have been various attempts to evolve the vertically integrated, technology-specific OSS/BSS software architecture to a vertically disintegrated, layered architecture, where a single piece of software takes care of the process regardless of the network technology used. Use of single technology-independent systems for customer-related processes (e.g., CRM) has achieved this best with the aid of generic software vendors adapting their horizontal (i.e., industry independent) software for the requirements of CSPs. At the service layer the attempt was known as IMS. Now service delivery platform (SDP) or frameworks (SDF) try to bring together all the interfaces needed for developing a new service in a single, standardized environment. In practice, these SDP/SDF approaches are merely enforcing an engineering practice making such a work easier rather than establishing an engineering environment or tool for the automation of new service creation and provisioning.

The next approach to standardize interfaces for OSS/BSS software is to standardize the network interface into a single abstraction, which should then be able to interoperate with various wireless and fixed networks with different performance characteristics. The most often cited approach for this is applying all-IP and internet protocols also to wireless mobile communication using new protocols for solving technical problems related to roaming and identity management. The main challenges CSPs seem to face with this approach come from the decreasing value of prior investment on old technologies, from the migration costs as well as from uncertainty of new business models. As billing basic voice communication becomes impossible, the revenue should come from monthly fees, added value services, advertisements and other new sources.

All these changes related to all-IP adoption will have a major impact on OSS/BSS software. Billing systems accounting for a half of the OSS/BSS product revenue will become less necessary if monthly flat fees are commonly adopted. OSS/BSS software for customer identity management relying on phone number or SIM card identity will have to be revised, etc.

If advertisements were to become the major income for consumer communication service providers, the operator segment might evolve towards the dual model of (1) bit-pipe operators servicing both enterprise customers and advertisement agencies subsidizing the monthly fees of consumers, and (2) specialized value added service operators providing service packages or individual services for the packagers.

Yet another possible future for CSPs is that they specialize more and more on developing new value added services for customers. In this scenario the operators still own networks, although they have outsourced network operations to third parties, either to global IT integrators using their internal OSS/BSS software, or to network equipment vendors with their proprietary OSS/BSS systems. Later on, CSP operations outsourcing may evolve into white-label bit-pipe operators servicing multiple CSPs with their gradually extending networks.

In most of these scenarios the number of OSS/BSS software vendors providing software related to the network layer is likely to decrease along with the decreased number of network equipment vendors. This might be the case even if the introduction of new network technologies still remained active. As the analysis of mergers and acquisitions show, the number of software vendors seems to be decreasing although the vertical software market has not matured yet. As a general implication, the industry evolution life cycle and the vertical software market evolution seem to be only loosely coupled and the lack of software market growth causes market consolidation even in the middle of the vertical software market evolution. This would imply that the vertical software market evolution has stopped to the second phase from the network technology viewpoint. However, the number of software vendors focusing on the customer interface layer is more difficult to predict as the border line between CSP software providers and the vast number of internet service software providers becomes more blurred due to the market convergence.

3 Future Work on Vertical Software Industry Analysis

This analysis of telecom operator software industry was based on the previous research of vertical software markets in Finland (Tyrväinen et al. 2004) and on an analysis of six vertical software industries: construction, energy, finance, forest, machine construction, and retail (Tyrväinen et al. 2005). Focusing on one industry enabled gaining deep insights of the vertical software industry dynamics.

The future work will continue along multiple tracks. Within the OSS/BSS industry there are multiple targets for further analysis based on the changing business environment and the rapid technology development in the domain. These include topics related to the technology layer starting from analyzing the impacts of new network technology to the software vendors, adoption of all-IP technology through dominating LTE or WIMAX technologies as well as analyzing approaches where network management and even the whole network layer is increasingly acquired as managed services.

In the service layer, potential research targets include analysis of adoption of service development platforms which can potentially form a dominant design for the industry. Most interestingly, the new business models in the user layer can revise the business logic of the consumer segment of the telecom industry with, e.g., advertisement-driven business models. During this research software as a service (SaaS) has grown into a relatively large market segment. There exists a natural synergy in CSP business and SaaS, which makes analysis of alternative value networks enabled by SaaS very useful though challenging.

Finally, the insights gained in the analysis of the OSS/BSS industry provide a fruitful basis for analyzing other vertical software markets as well as for cross-industry comparisons based on indicators. Through this work a deeper understanding on the dynamics of vertical software industries will be gained.

References

Agarwal R (1998) Evolutionary trends of industry variables. Int J Ind Organ 16(4):511–525

Argawal R, Audretsch DB (1999) The two views of small industry dynamics: a reconciliation. Econ Lett 62(2):245–251

Ali-Yrkkö J and Martikainen O (2008) The Impact Of Technological And Non-Technological Innovations On Firm Growth. Helsinki, ETLA, The Research Institute of the Finnish Economy, p. 19 (Keskusteluaiheita, Discussion Papers; ISSN 0781-6847; no. 1165)

Altenberg L (1995) Genome growth and the evolution of the genotype-phenotype map. In: Banzhaf W, Eckman FH (eds) Evolution and biocomputation. Springer, Berlin, pp 205–259

Ansoff HI (1987) The emerging paradigm of strategic behavior. Strategic Manage J 8:501–518

Barney JB (1991) Firm resources and sustained competitive advantage. J Manage 17:99–120

Bass F (1969) A new product growth model for consumer durables. Manage Sci 15:215–225

Bekkers R and West J (2006) The effect of strategic patenting on cumulative innovation in UMTS standardization, DIME Working Paper, No. 8

Bell J (1995) "The internationalization of small computer software firms A further challenge to "stage" theories". Eur J Mark 29(8):60–75

BMBF (2000) Analyse und Evaluation der Softwareentwicklung in Deutschland. Bundesministerium für Bildung und Forschung, German

Bower JL and Christensen CM (1995) "Disruptive technologies: catching the wave," HBR

Bresnahan TF (1998) "New modes of competition and the future structure of the computer industry," in Competition, Convergence, and the Microsoft Monopoly. Kluwer Press, New York

Bresnahan TF, Greenstein S (1999) Technological competition and the structure of the computer industry. J Ind Econ 47(1):1–40

Campbell-Kelly M (2004) From airline reservations to sonic the hedgehog. MIT Press, Cambridge

Chesbrough H, Rosenbloom R (2002) The role of the business model in capturing value from innovation: evidence from Xerox Corporation´s technology spinn-off companies. Ind Corp Change 11(3):529–555

China Unicom (2008) Official site, accessed on Nov. 12, 2008 at http://www.chinaunicom.com/english

Church J, Gandal N (1993) Complementary network effects and technological adoption. Int J Ind Organ 11:239–260

Churchill NC, Lewis VL (1983) The five stages of small business growth. HBR 61(3):30–49

Colecchia A, Anton-Zabalza E, Devlin A, Montagnier P (2002) Measuring the information economy. OECD Organization for Economic Co-operation and Development, Directorate for Science, Technology and Industry, Paris, France

Cooper R, Zmud R (1990) Information technology implementation research: a technological diffusion approach. Manage Sci 36(2):123–139

Del Monte A, Papagni E (2003) R&D and the growth of firms: empirical analysis of a panel of Italian firms. Res Pol 32:1003–1014

Dinlersoz E, Hernández-Murillo R (2005) The diffusion of electronic business in the United States. Fed Reserve Bank St Louis Rev 87(1):11–34

Economides N (1991) "Compatibility and the Creation of Shared Networks."

EITO European Information Technology Observatory (2004)

ETSI SR 000 314 V1.15.1 (2005) Special Report, Intellectual Property Rights (IPRs); Essential, or potentially Essential, IPRs notified to ETSI in respect of ETSI standards, p. 2751

Fontana A, Frey JH (2000) The interview: from structured questions to negotiated text. In: Denziu NK, Lincoln YS (eds) Handbook of qualitative research. London, Sage, pp 645–672

Frakes W, Terry C (1996) 'Software reuse: metrics and models,'. ACM Comput Surv 28(2):415–435

Frank L (2004) An analysis of the effect of the economic situation on modeling and forecasting the diffusion of wireless communications in Finland. Technol Forecast Soc 71:391–403

Frank L, Luoma E and Tyrväinen P (2007) "Market scope of vendors in the OSS software market," Proceedings of the IEEE International Conference on Industrial Engineering and Engineering Management (IEEM), pp 2096–2100

Frank L and Luoma E (2008) Telecommunications Software Industry Evolution, in proceedings of 7th Conference on Telecom, Internet & Media Techno-Economics (CTTE 2008), Paris, France, 18 June, 2008

Freeman C (1994) The economics of technical change. Cambridge J Econ 18:463–514

Frenken K (2006) A fitness landscape approach to technological complexity, modularity, and vertical disintegration. Struct Change Econ Dynam 17:288–305

Garud R, Kumaraswamy A (1993) Changing competitive dynamics in network industries: an exploration of Sun Microsystems' Open Systems strategy. Strategic Manage J 13(5):351–369

Giarratana M (2004) The birth of a new industry: entry by start-ups and the drivers of firm growth the case of encryption software. Res Pol 33:787–806

Gort M, Klepper S (1982) Time paths in the diffusion of product innovations. Econ J 92 (267):630–653

Grant RM (1996) Toward a knowledge-based theory of the firm. Strategic Manage J 17:109–122

Greiner LE (1972) Evolution and revolution as organizations grow. HBR 50(4):37–46

Griliches Z (1957) Hybrid corn: an exploration in the economics of technological change. Econometrica 25:501–522

Gruber H, Verboven F (2001a) The diffusion of mobile telecommunications services in the European Union. Eur Econ Rev 45:577–588

Gruber H, Verboven F (2001b) The evolution of markets under entry and standards regulation – the case of global mobile telecommunications. Int J Ind Organ 19:1189–1212

Harrison R, Jaumandreu J, Mairesse J, Peters B (2008) Does innovation stimulate employment? a firm-level analysis using comparable micro-data from four European countries. NBER Working Paper 14216, Cambridge MA, pp 1–46

Hoch DJ, Roeding CR, Purkert G, Lindner SK, Müller R (2000) Secrets of software success. Harvard Business School Press, Boston

Katz ML, Shapiro C (1985) Network effects, competition, and compatibility. Am Econ Rev 75:424–440

Mansfield E (1961) Technical change and the rate of imitation. Econometrica 29:741–766

ITU-T Recommendation M.3010 (2000) Principles for a telecommunications management network. International Telecommunication Union

Jeung F (2008a) Beijing confirms 3G licences to be issued. South China Morning Post, May 26, 2008. Accessed on Nov. 11, 2008 at www.scmp.com ->archive

Jeung F (2008b) China Unicom records 34.8pc rise in net profit South China Morning Post, Oct. 31, 2008. Accessed on Nov. 11, 2008 at www.scmp.com ->archive

Jeung F (2008c) China Unicom studying integration of operations South China Morning Post, Nov. 5, 2008. Accessed on Nov. 11, 2008 at www.scmp.com ->archive

Jeung F (2008d) Let homegrown 3G compete in market, Qualcomm urges. South China Morning Post, Oct. 28, 2008. Accessed on Nov. 12, 2008 at www.scmp.com ->archive

Jeung F (2008e) Telecoms regulators take aim at cut-throat rivalry South China Morning Post, Nov. 11, 2008. Accessed on Nov. 12, 2008 at www.scmp.com ->archive

Jeung F (2009a) Telecoms stocks fall as 3G licences are issued. South-China Morning Post, January 8, 2009. Online at http://archive.scmp.com/gotoshowarticles.php?artid=20090108946610360

Kazanjian RK (1988) Relation of dominant problems to stages of growth in technology-based new ventures. Acad Manage J 31(2):257–279

Kelly M (2003) "The Tele management forum's enhanced telecom operations map (eTOM)". J Netw Syst Manage 11(1):109–119

Kottler P, Keller Lane K (2005) Marketing management. Prentice Hall, New York

Kuznets S (1930) Secular movements in production and prices. Houghton Mifflin, Boston

Lewin K (1947) Group decision and social change. In: Newcomb TM, Hartley EL (eds) Readings in social psychology. Henry Holt, New York

Macher JT, Mowery DC (2004) Vertical specialization and industry structure in high technology industries. Adv Strategic Manage 21:317–356

Maliranta M and Rouvinen P (2003) Productivity effects of ICT in Finnish business, ETLA Discussion Paper 852

Maliranta M and Rouvinen P (2004) Informational mobility and productivity – Finnish evidence. ETLA Discussion Paper 919

Mazhelis O, Tyrväinen P, and Viitala E (2007) "Modelling Software Integration Scenarios for Telecommunications Operations Sotware Vendors," Proceedings of the IEEE International Conference on Industrial Engineering and Engineering Management (IEEM), December 2–5, Singapore

Mazhelis O, Tyrväinen P, Frank L and Suojapelto K (2008a) Analyzing vertical software market evolution: the Case of Telecom OSS/BSS software. Proceedings of the Next Generation Communications/Public-Private Interplay Conference, Sevilla, Spain December 10–12

Mazhelis O, Tyrväinen P, and Matilainen J (2008b) Analyzing Impact of Interface Implementation Efforts on the Structure of a Software Market: OSS/BSS Market Polarization Scenario, in Proc. of the 3rd International Conference on Software and Data Technologies (ICSOFT 2008), Porto, Portugal, 5–8 July

Mazhelis O, Tyrväinen P, Viitala E (2008c) Analysing software integration scenarios: the case of telecommunications operations software. Int J Manage Sci Eng Manage 3(3):200–210

Mazhelis O, Tyrväinen P, Frank L and Viitala E (2008d) Modeling the impact of software development and integration costs on the software market size. In the proceedings of the WWW Internet conference, University of Freiburg, Germany 13.–15.10.2008. Institute for the Management of Information Systems, IMIS

McHugh P (1999) Making it big in software. Rubiv Publishing, Triverton

Moore GA (1995) Inside the tornado: marketing strategies from Silicon Valley's cutting edge. HarperCollins Publishers, New York

Morone P, Testa G (2008) Firms growth, size and innovation An investigation into the Italian manufacturing sector. Econ Innov New Techn 17(4):311–329

Mueller D (1980) The Determinants and effects of mergers: international comparison. Cambridge University Press, Cambridge

Murmann JP, Frenken K (2006) Toward a systematic framework for research on dominant designs, technological innovations, and industrial change. Res Pol 35(7):925–952

Nambisan S (2002) Software firm evolution and innovation-orientation. J Eng Technol Manage 19:141–165

Northrop L (2002) SEI's software product line tenets. IEEE Software 19(4):32–40

OSS/J Product Team (2006) "The OSS through Java™ API Roadmap. Version 3.1 (Public Final)". The OSS through Java™ Initiative

Porter ME (1985) Competitive advantage. The Free Press, New York, US

Poulin JS, Caruso JM, Hancock DR (1993) The business case for software reuse. IBM Syst J 32(4):567–594
Pulkkinen M, Frank L, Veijalainen J (2008) The e-Society carriers: developments with communication services providers (CSP's). Public/Private interplay in next generation communications, Seville, Spain December 10–12
Robson C (2002) Real world research. Blackwell Publishing, Malden, MA
Rogers EM (1983) Diffusion of innovations, 3rd edn. The Free Press, New York
Romijn H, Albaladejo M (2002) Determinants of innovation capability in small electronics and software firms in southeast England. Res Pol 31:1053–1067
Rust R, Lemon K, Zeithaml V (2004) Return on marketing: using customer equity to focus marketing strategy. J Mark 68:109–127
Ryan B, Gross NC (1943) The diffusion of hybrid seed corn in two Iowa communities. Rural Sociol 8:15–24
Sanz J, Nayak N and Becker V (2006) "Business services as a modeling approach for smart business networks," Technical paper, Available online
Sawhney M, Wollcott RC, Arroniz I (2006) The innovation radar: A 360-Degree view of business innovation. Sloan Manage Rev 47(3):75–81
Schoder D (2000) Forecasting the success of telecommunication services in the presence of network effects. Inform Econ Pol 12:181–200
Schoemaker PJH (1995) Scenario Planning a Tool for Strategic Thinking, Sloan Management Review, Winter pp. 25–40
Schumpeter J (1939) Business cycles. McGraw-Hill, New York
Scott M, Bruce R (1987) Five stages of growth in small businesses. Long range planning 20(2):45–52
Stigler GJ (1950) Monopoly and oligopoly by merger. Am Econ Rev 40:23–34
Stigler G (1951) The division of labor is limited by the extent of the market. J Pol Econ 59(3):185–193
Stigler GJ (1958) The economies of scale. J Law Econ 1:54–71
Terplan K (2001) OSS essentials – support system solutions for service providers. Wiley, New York, NY
Thomsen O (2006) The lean approach. Wireless Pers Commun 38:17–25
TMF GB921 (2005) Enhanced Telecom Operations Map (eTOM) – The business process framework, Version 6.1. Tele Management Forum, Morristown, NJ
Tushman ML, Anderson P (1986) Technological discontinuities and organizational environments. Adm Sci Q 31:439–465
Tyrväinen P, Frank L (2008) Diffusion of telecom software services and products - from legacy services to focused products, July 22–27. IADIS Telecommunications, Networks and Systems, Amsterdam, The Netherlands
Tyrväinen P, Lamberg I, Nukari J, Saukkonen J, Seppänen V, ja Warsta J (2004) "Innovatiivisilla toimialaratkaisuilla kansainvälisille ohjelmistomarkkinoille," TEKES Teknologiakatsaus 178/2005, ISSN 1239-788X, ISBN 952-457-206-0, Helsinki 78 + 58 p
Tyrväinen P, Warsta J, Seppänen V (2008) "Evolution of Secondary Software Businesses: Understanding Industry Dynamics". In: León G, Bernardos A, Casar J, Kautz K, DeGross J (eds) IFIP International Federation for Information Processing, Vol. 287, Open IT-based innovation: moving towards cooperative IT transfer and knowledge diffusion. Springer, Boston, pp 381–401
Tyrväinen P, Warsta J, and Seppänen V (2004) "Toimialakehitys ohjelmistoteollisuuden vauhdittajana - Uutta liiketoimintaa lähialoilta," TEKES Teknologiakatsaus 151/2004, ISSN 1239-788X, ISBN 952-457-143-9, Helsinki, p. 71
Valente TW (1995) Network models of the diffusion of innovations. Hampton Press, Cresskill, NJ
Wang Z (2006) "Learning, diffusion and industry life cycle". Payments System Research Working Paper, Federal Reserve Bank of Kansas City
Wernerfelt B (1984) The resource-based view of the firm. Strategic Manage J 5:171–180

Williamson OE (1975) Markets and hierarchy: analysis and antitrust implications. Free Press, New York

Witt U (1997) ""Lock-in" vs. "Critical masses" - Industrial change under network effects". Int J Ind Organ 15:753–773

Yang C-H, Huang C-H (2005) R&D, size and firm growth in Taiwan's electronics industry. Small Bus Econ 25:477–487

Xinhua (2008) China Netcom, China Unicom merger completed, biggest in country's history. Xinhua Oct. 15, 2008. Accessed on Nov. 12, 2008 at http://news.xinhuanet.com/english/2008-10/15/content_10200183.htm

Yasuda T (2005) Firm growth, size, age and behaviour in Japanese manufacturing. Small Bus Econ 24:1–15

Index

I

K

L

M

N

O

P

FSC
www.fsc.org
MIX
Papier | Fördert gute Waldnutzung
FSC® C083411

Kolibri 360 GmbH
Ferdinand-Jühlke-Straße 7
99095 Erfurt, Deutschland
produktsicherheit@kolibri360.de